D0305943

DEAR NHS

100 STORIES TO SAY THANK YOU

EDITED BY

ADAM KAY

TRAPEZE

First published in Great Britain in 2020 by Trapeze,
an imprint of The Orion Publishing Group Ltd
Carmelite House, 50 Victoria Embankment, London EC4Y 0DZ

An Hachette UK company

5 7 9 10 8 6 4

A CIP catalogue record for this book is available from the British Library.

ISBN (Hardback): 978 1 398 70118 2
ISBN (eBook): 978 1 398 70137 3

Typeset by Born Group
Printed and bound in Great Britain by Clays Ltd, Elcograf, S.p.A

www.orionbooks.co.uk

CONTENTS

INTRODUCTION

I'm very proud to share with you *Dear NHS: 100 Stories to Say Thank You*. In the pages that follow, 109 well-known people recount their personal experiences of the health service. Yeah, about that. It's not a typo. One hundred and nine. Asking people to be involved was a bit like posting out invites to a wedding: you send a few too many because you assume there'll be a bunch of people who can't make it. Well, practically everyone said yes and we'd already announced the title.[1]

This is slightly embarrassing because my role as editor involved two simple tasks – assembling 100 stories and writing 500 words of introduction (I went over on that one too). I didn't even have to edit out spelling mistakes – the publishers did that for me.[2]

1 I'll probably tell you the people who said no if you buy me a drink.

2 Incidentally, the group of people who made the most spelling mistakes were the professional writers. Absolute shambles.

Unlike when my wedding proved more popular than expected, however, I couldn't just shove an extra table in the corner of the marquee for the least important guests. (Apologies to any of my family who read this.) My first thought was to play bouncer in a shit nightclub and do one-in-one-out, but that felt slightly unfair. Then I realised that if I could count both Hairy Bikers as one entry, I could attempt to persuade you that there's a pop group comprised of Ian Rankin, Emma Watson, Trevor McDonald, Lorraine Kelly, Joanna Lumley, Johnny Vegas and Malala.[3] Instead, I approached it as I would a 4 a.m. kebab and just crammed the lot in. The fact that so many people said yes is simply testament to the love we all rightly have for the NHS: a love we inherit from our parents and which burns brighter with every hospital appointment. Whoever we are, however famous, we've all been touched by the health service and we all want to say thank you. Now more so than ever.

The stories in this book aren't just testimonials; they are memories relived, secrets shared; the comedies and tragedies of everyday life that we all recognise. They're by turns funny and heartbreaking, uplifting and moving, and all deeply personal and utterly heartfelt. They were also a stark reminder to me that they never teach you at medical school what to say when you see a famous patient. Is it more professional to pretend you don't recognise them or to gush that you simply *loved* them in that film?

3 Tell me you wouldn't buy a ticket?

In my first year as a doctor, I found myself checking over a singer (A-list) who was on tour in this country and had a funny turn with no identifiable cause whatsoever, save for the enormous quantity of drugs he had just consumed (A-class). When I asked his occupation, he just raised an eyebrow at me that said, 'You know who I am.'[4] I still slightly regret not having my stethoscope signed. Anyway, as I read the star-studded submissions to this book, I couldn't help but put myself in the doctor's shoes and wonder what the hell I'd say as I, for example, extracted a hoover attachment from within Sue Perkins or deflated Louis Theroux's swollen testicle.[5]

It means a lot that you've bought a copy of this book. I'm so grateful to you. As well as saying thank you in one hundred(ish) separate ways, *Dear NHS* also says thank you in a very practical way, by raising money for NHS Charities Together and The Lullaby Trust. Now that Captain Tom has set the fundraising bar fairly high, you might need to buy a few more copies, to be honest.

I'm also hugely grateful to every single contributor for their generosity – not only in giving up time to write their chapters but for opening up so honestly about often deeply personal and painful matters.

But most of all, I'm grateful beyond words to the NHS. To all the doctors and nurses and midwives and

4 It'll cost you three drinks minimum to get this person's name.
5 Spoiler alert, sorry.

paramedics and pharmacists and physios and OTs and ODPs and optometrists and carers and speech therapists and radiographers and cleaners and social workers and dieticians and health visitors and admin teams and district nurses and porters and podiatrists and managers and kitchen staff and healthcare assistants and ward clerks and biomedical scientists: thank you. The NHS is our single greatest achievement as a country, and the NHS is *you*. The 1.5 million people who go above and beyond the call of duty every single day. The ones who give us hope and make sure there's a tomorrow. 1.5 million people for whom the extra mile is the standard distance. Selflessly, generously putting others before yourselves. You've been there for me so many times, you've saved the lives of many people I love and you've done the same for every single person reading this book. This is our way of saying thank you.

Adam Kay

GRAHAM NORTON

I avoid the news. Stay Home. Protect the NHS. Save Lives. This is all I need to know. Anything else just increases the mysterious weight pressing against my chest. An unnamed dread, a sense of doom. Even the bright spring weather, normally so cheering and welcome, doesn't help. I feel as if we have laid out a beautifully prepared banquet but inadvertently used the wonky trestle table, the one that collapses if you put any weight on it. We are all staring helplessly at quivering jellies and glistening bowls of coronation chicken, hoping against hope that the table holds.

I can't avoid the news. I wake to see a tweet from a nurse. She describes her day at work and how she had to hold up a phone to the ear of not one, not two, but four patients, so that their loved ones could say goodbye. I begin my day in tears. A virus so cruel that it doesn't just kill, but first finds brutal and unexpected ways of punishing people. No doctor, no nurse is trained for this.

I find it impossible to imagine what it must be like for those working on the front line. Happily, my experience of hospitals, thus far, is very limited. I've visited friends, of course, and surely in the thirty-six years I've lived in this country I must have sat in an A&E department at some point, but I genuinely can't recall doing so. My only extended visit to hospital took place in the summer of 1989.

There had been a party at my drama school in Swiss Cottage in North London. Afterwards, the drunk me had a choice: get a taxi home or walk and spend the taxi fare on fried chicken. Chicken won the day. I stumbled down the hill towards Kilburn and then headed across to Queen's Park, where I had a flat. Licking my greasy fingers, I didn't notice the point at which I began to be followed.

I was very close to home, on the street that edged the north side of the park, when I noticed a man walking on the pavement opposite me. He crossed the road and began walking in front of me. Suddenly he turned and seemed to be brandishing something. A bat, a bar? I wasn't sure.

Turning to get away, I ran straight into his accomplice, who had been right behind me. I heard a strange hollow banging sound, like someone hitting a large plastic pipe. It was only afterwards I realised that was the sound of my skull being beaten. After I handed over my empty wallet, they made me lie on the ground as they emptied my rucksack. A dog-eared copy of *The Winter's Tale* flopped in the gutter beside me.

They ordered me to stay on the ground as they made their getaway. I heard their footsteps echo as they ran away through the deserted streets of Queen's Park. As I waited for them to leave, I noticed a cut on my right hand. They must have had a knife. How strange that I hadn't noticed. Oh well, it was just a nick.

I pushed myself off the ground to stand. Odd. It felt as if I was almost peeling my T-shirt off the pavement. I looked down to discover that I was covered in blood. I pulled my top away from my body and saw that I had a hole in my chest. Even in my drunken state I knew that this was serious. I collected my belongings and put them back in my bag, because it takes a moment to understand what matters. I felt very tired. I wanted to lie down. That would be a bad idea.

I stumbled on, calling for help. The houses that lined the leafy street were in darkness. I rang a doorbell, but no one came. Back on the street I walked a little further and then, to my right, there was a light. An old man in a dressing gown stood framed in his front doorway. His wife was huddled behind him. I stood at their gate. Obviously an explanation was required. I lifted up my T-shirt and pointed at the open wound in my chest. 'I've been stabbed.' It had echoes of 'I am run through' from Shakespeare or John Lennon's 'I've been shot.' Something so obvious, but it still needs to be articulated because it is so surprising.

I am not sure if the elderly couple said anything, but I felt I had been rescued. I walked down the tiled garden path and lay bleeding on their doormat. The man must have gone to phone an ambulance because I recall being left alone with the old lady. Weariness overtook me. All I wanted to do was sleep. No. That wasn't true. There was one more thing I needed. I looked up at the lady in her dressing gown and asked, 'Will you hold my hand?' Clearly taken aback by the request, she hesitated, before kneeling down and taking my hand. She was the nurse in that tweet holding the phone up. The contact that meant I was not alone, at that moment when nobody should be alone.

I was taken to the St Charles Hospital in Ladbroke Grove. It transpired that I had lost over half my blood. Even after I was informed of this, I still didn't fully understand the seriousness of the situation. I was still the boy in the street with a hole in his chest, stopping to gather up his books. It was only when a nurse asked me if I wanted the hospital to contact my parents that I got an inkling of how touch-and-go things were. I thought about her question. I didn't want to worry them unnecessarily, but equally I knew how annoyed they'd be if they didn't get to say goodbye, so I simply asked the nurse, 'Am I going to die?' The long pause before she gave her uncertain response, 'No,' made my flesh hug my bones.

The recovery from a violent mugging takes a long time. It's not just physical, but also mental. Your lungs regain their strength long before you stop flinching when a stranger gets too close on the street. Oddly, the very existence of the NHS helps.

My father would never have said such a thing, but he must have felt vindicated by what had happened. London was a very dangerous place and so of course I had been stabbed. He couldn't understand why I would stay in such a death trap. I was one of those farmers who continue to live and work on Mount Etna. What I found hard to explain to him was how my two-week stay in hospital hadn't made me more fearful about life, it had reassured me, made me feel safer. When I was a boy learning to ride a two-wheeler, I had been frightened and excited, but I knew that right by my side was my father waiting to catch me if I fell. That is living in the UK with the NHS. They are always there to catch us.

As I write this, I have no idea how long our current situation will last, or what life will be like in the world that comes after this, but of one thing I am confident. We will gather together again, pass plates, break bread, raise glasses. The feast will still be there to be enjoyed, because, despite our fear and our doubts, the table will hold.

LEE MACK

You Can Leave Your Hat On

'Is it socially acceptable to leave your hat on when someone is sticking their finger up your bum?' is a question all of us never ask ourselves on a daily basis. But it's a question I once had to ask on a visit to an NHS doctor.

This story starts many years ago when I was in the ITV Christmas panto with *One Foot in the Grave*'s Richard Wilson. We were having drinks in the pub after rehearsal and I noticed that every time it was Richard's turn to go to the bar he would put a baseball cap on his head. Then, when he returned, he would take it off. He explained that the wearing of a hat was extremely effective for anonymity and the difference between people shouting his catchphrase at him and not. I told him I didn't believe it. He didn't laugh.

Cut to many years later and I had now started appearing more on television myself, so I too had started getting recognised. Although I sometimes wonder if they were mixing me up with someone who had the catchphrase,

'How on earth do you get away with it, you chancer?'

Now, to be clear, my attitude towards being recognised has always been, on the whole, that it's perfectly fine. Occasionally it can be a pain in the arse (we'll be getting to the NHS doctor bit in a moment), especially when, for the fiftieth time that week, somebody says, 'I thought you didn't go out,' which is obviously a comment about my sitcom *Not Going Out*. In fact, if in 2006 I could have magically predicted the coronavirus lockdown fourteen years later and the following torrent of online jokes, perhaps I'd never have called my sitcom *Not Going Out* in the first place.

Because I don't consider myself household famous like, say, Piers Morgan, Simon Cowell or Dennis Nilsen, being recognised is not so regular an occurrence that I have strong feelings about it either way, but there are definitely times I want my anonymity. And for some reason, a hat is the most effective way of achieving that. I think it's something to do with changing the shape of your face. Or perhaps it's more to do with the fact that people are saying, 'A man wearing a fez in Tesco is very odd, so keep your distance, kids.'

And one of the times that sort of anonymity is important is when you're having your first prostate examination with an NHS doctor.

I was going to see the doctor about an altogether different matter, although I can't remember what that was. And I *honestly* can't remember. I don't mean I *do* remember but

I don't want to tell you because it's embarrassing. Because of course, I am a stand-up comedian. And, as we all know, stand-up comedians would happily get erectile dysfunction if it meant they had something new to talk about at this year's Edinburgh Fringe. Although I guess erectile dysfunction would be quite an ironic subject for stand-up.

I was about to turn fifty and I kept being told that is the age when you need to ask the doctor to stick his finger up your bum and check for anything dodgy. They are not the exact words you use, and it is usually him that asks you, so probably best you don't walk in to the GP's surgery and quote that verbatim. At the very least, say hello first.

So, I had decided that, while visiting the doctor for something that definitely wasn't erectile dysfunction, I was going to bring up the subject of the prostate exam.

Sitting in a doctor's waiting room is a perfect example of what I would class as 'anonymity important'. The thought of other people thinking, 'Oh, it's that bloke from Eight Out of Ten Ways to Mock the Week. I wonder what he's here for? I bet it's a rash. Or erectile dysfunction.' Have I perhaps mentioned erectile dysfunction too many times? Doth the lady protest too much? To be clear, I absolutely one hundred per cent promise I am not a lady with erectile dysfunction.

I had to wait quite a time in that waiting room. The clue's in the name, I guess. During this time, I started having the most intense internal debate in my head about what would happen once in the room with the doctor. What if the doctor said, 'I'll do the prostate examination right now'? I was nervous; I'd never had one before. And I would like to pretend that I am the kind of mature person who thinks, 'It's nothing to be embarrassed about. This is an important thing to do.' But I'm not that person. I am a person so incredibly immature that I have actually made a living out of it. So instead I got into a right tizz about the whole thing.

And during this tizz (I have literally *never* used the word 'tizz' in my life and I have now just used it three times in two sentences. I think it might be PTSD brought on by reliving the incident), I decided that, once in the room with the doctor, it was probably best that I leave my hat on in case the prostate examination happened. It seemed to make sense to me. If Richard Wilson felt it appropriate to keep his hat on to stay anonymous when buying a drink, then *surely* it was fair enough to keep my anonymity when somebody I have never met has a part of his anatomy in mine.

But then the social etiquette pendulum started swinging massively. For anonymity reasons, the one time you definitely want to keep your hat *on* is when someone has his finger up your bum, but it can also be argued that the one time you should definitely take your hat *off* is when someone has his finger up your bum. If social etiquette forces you to remove your hat in a place of worship, then surely those rules apply to a prostate exam. I mean remove your hat, I don't mean you should never have a prostate exam in a place of worship. Although you shouldn't.

I knew in my heart of hearts it wasn't a debate. Surely the right thing to do was 'hat off'. Especially given it wasn't a baseball cap, which is a bit of headwear you see people wearing all the time. Mine was more of a fedora shape. And if you don't know your hats, then imagine a 1950s detective about to solve a case. Now imagine him with his trousers round his ankles and someone putting their finger up his bum.

Of course, there was no guarantee that the hat would work. What if he *did* recognise me? That is then surely ten times worse. I imagined him sitting with his family over dinner saying, 'I had that bloke in today from Eight Out of Ten Ways to Mock the Week.' 'Oh yeah, what was he like?' 'He asked for a prostate examination and guess what? He left his bloody hat on!' 'Wanker. I've never liked him.'

So, what did I do? Well, I'm afraid this story ends rather flatly. I did decide I would keep my hat on, but the prostate examination never happened. Well, not on that day anyway. When it did happen, to avoid the same internal debate, I decided not wear a hat at all.

Which I know was the right thing to do, but also a bit of a shame. One of my comedy heroes is Stan Laurel. Part of me wanted to learn that thing that he used to be able to do, which was to make his hat rise up in the air slightly, seemingly on its own. If I could have done that at the exact moment the doctor's finger went in, it would have been my greatest comedic achievement ever.

See, I told you I was immature.

DAME EMMA THOMPSON

This is a memory of an NHS experience that I will never forget. It was the middle of the night and suddenly the light was on and my eighty-five-year-old mother, who suffers from Parkinson's, was standing in the doorway of our bedroom in her nightie, which was on back to front.

'Oh God,' I thought. 'She's died and has come to tell me herself.'

Which would have been typical and hard enough but no, she spoke. 'Gaia's in hospital,' she said.

Gaia, our daughter, was eighteen at the time.

Greg, my husband, who has the aural capacity of an anxious bat, leapt from the bed – naked – to grab his clothes. My mother shrieked at the sight of his flailing gonads approaching at speed. I too leapt from the bed and fell over my clothes, bashing my head on the wardrobe door, which I had ill-advisedly left open. It was suddenly a scene from a Brian Rix farce but performed in an experimental style by French actors.

We put my mum, still trembling, back to bed and drove at an illegal speed to the Whittington Hospital in Archway, London. The A&E was quiet, no one was screaming and there was no blood on the floor. Our daughter was sitting in bed surrounded by five girls all laughing like drains. She informed us with great brio and apparently no pain that she had slipped over at a friend's party and opened her head on a sharp bit of tile.

The junior doctor in charge was very young and very kind. Gaia told me that when she arrived, brought in by my friend whose son's party was still in progress, the doctor had – very apologetically – asked if it was all right if he just quickly went to eat something as he was too hungry to concentrate. But by the time we had arrived, he was back from swallowing a sandwich – and very quick to do what was necessary.

He injected Gaia with a local anaesthetic and allowed me to hold the pieces of skin together while he stitched her up. Gaia was immensely cheerful – probably owing to the alcohol that had led to the fall in the first place. I'll never forget that doctor – his kindness, his patient care and, most of all, his infinite fatigue. I thought, this is all wrong. People doing this kind of vital work shouldn't have to be at it so long that they get this tired and too overworked to eat. Why do we do this to our doctors and our nurses?

In the A&E there was also a man of about fifty who too had hurt his head. He was very anxious about his mother – he was her principal carer and had no one to turn to in his hour of need. We offered to help but there was nothing we could do, he said. He just needed to get treatment and get back to her as soon as he could. There would certainly be no one but the NHS to look after him.

You've heard of The Untouchables. You've heard of The Expendables. That doctor, that hurt carer, belong to The Indispensables. Right now it is no exaggeration to say that they are the only ones standing between us and complete calamity. Who else, doing what job and in what capacity, can make that claim?

Four weeks into lockdown and I've listened daily to the stories of heroism and sacrifice and unimaginable loss that occur every moment of every day and every night within the NHS – and in hospices and care homes and ordinary homes. I marvel at the courage; I weep for the losses and I know that after this we will never be the same again. At least, I hope we won't. I hope we will realise what is important: not profit but people. I hope that every priority we have had to accept as 'normal' changes and that we find our way towards a society that cares first and foremost about its indispensable workers.

First of all, let's stop pretending that there's no need to pay them generously because caring professions are 'vocational'.

We can clap and we can whoop and holler our gratitude from the rooftops, and we should, and we do – but after this we all have to insist that whatever government is in place takes immediate action to recognise The Indispensables by raising their wages. By listening to what has been and is being learned, and what we now know is essential to have in place for the next time this happens – which, the science seems to suggest, is inevitable.

Our NHS was created after a great crisis – a world war. After this crisis must come a great and deep reckoning: who and what do we really value and how do we protect, nurture and support them.

LOUIS THEROUX

My Testicle

It was late 2015. I'd recently completed a film about the mysterious and secretive and supposedly highly litigious religion of Scientology. But the film did not yet have a distributor. In order to drum up interest, we had shown it at the London Film Festival. I'd done my best to support the premiere by doing interviews, and writing articles, and sending tweets.

One evening, a message came in from Simon, the film's producer. Something in his tone made me think it was ominous and I called back, after the kids' supper, from the quiet retreat of our top-floor bedroom, to hear him explain that he'd had a letter from lawyers acting for the leader of Scientology, David Miscavige. He said: 'Apparently you sent a tweet that they consider libellous and they are threatening legal action.'

'Oh dear,' I said.

'Yeah.'

'That's not good, is it?'

'No. Sorry.' This was said in a manner so heartfelt and final that it suggested not just that the threat really was serious, but also that there wasn't much he would be able to offer in the way of help. It was my Twitter account and if I was going to use it to libel vengeful high-profile figures, that was on me.

He ended by suggesting I call Nigel, the media lawyer who'd worked on the movie, which I did. He also forwarded me the letter. It quoted from a tweet – or rather a retweet, since the words had auto-generated when I'd clicked on a button to share an article – that said, and I quote, 'David Miscavige is a terrorist.' Yeah. That wasn't good. I recalled tweeting the article – I'd had some misgivings on account of its overheated content and had wondered about erasing it, but a publicist we'd retained had suggested I didn't, since apparently that was seen in PR circles as a sign of weakness and would probably only bring more attention. What I didn't recall – and didn't think I'd ever actually read – was the wording of the tweet itself. *David Miscavige is a terrorist.* I pondered all this a little ruefully, then called my agent. In a tone not dissimilar to Simon's, she said, 'David Miscavige may be a lot of things but he ain't a terrorist.'

'But what do I do?'

'You need to get a good lawyer and get ready to spend a lot of money. Because I'm telling you now, this could be very expensive.'

A day or two later, another letter arrived from the lawyers acting for Miscavige, filled with more legal sabre-rattling and shield-clanking – 'false', 'outrageous', 'defamatory' – and a demand for an apology. This I might have thought about providing – given that I don't actually think David Miscavige is a terrorist – except my lawyer warned that the apology would not forestall a claim of financial damages but, in fact, only make one more likely, possibly to the tune of £100,000 or more.

My lawyer advised me to instruct a high-powered QC. A name was suggested; Heather Rogers. She had once been part of the legal team defending the writer Deborah Lipstadt in her libel defence against the historian David Irving, whom she had labelled a Holocaust denier. There were meetings – the QC was as impressive as I'd expected – and, as she did her preparation, read the letters, read my tweets, the articles, and viewed the film, and as the bills came in and money haemorrhaged out, I found myself mainly reassured by her level of competence and only slightly distraught at the strangeness of having to pay someone hundreds of pounds to watch a film you've made, as research.

Around the same time, I was making trips up to MediaCity in Salford where I was appearing on a Christmas edition of the quiz show *University Challenge*. On the train, I would brood about my own stupidity at sending the tweet and the likelihood of it having catastrophic consequences. I looked up the meaning of 'terrorist'. You could 'terrorise' someone without doing them physical harm, I reasoned. Though, as the Scientology letters pointed out, my tweet had gone out not long after the Charlie Hebdo murders, so I was sort of suggesting that David Miscavige went round stabbing journalists, which he hasn't done as far as I know.

I did a Twitter inventory to see how many of my followers were real people. It suggested I'd only published the tweet to a million people, not 1.8 million. And in fact only a few thousand had probably seen it. I told this to Nigel, the lawyer.

'Yes, I'm not sure how helpful that is for us,' he said.

The case motivated me to do well on *University Challenge*. I was getting a small fee for each appearance. If my team went all the way, I'd only need another £98,000 for the war chest, though, come to think of it, that wasn't counting legal fees.

In the final I got on a hot streak, answering questions on *Mad Men*, Tennyson and Pope Linus I. We won. It improved my frame of mind for about fifteen minutes. Another legal letter came in from Miscavige's lawyers. We sent one back. Despite all the polysyllables and legal verbiage, it was, I realised, just a more sophisticated and more expensive version of two kids in a playground saying 'Come on then! If you want some!' 'You and whose army! Hold me back! Hold me back!' but neither of them really wanting to fight.

Still, it was stressful and not helping my equanimity was the sudden onset of a debilitating pain in the groin and, after ignoring it for a couple of days and then finally checking myself manually and in the mirror, the realisation that one of my testicles had grown to roughly four times its normal size. I racked my brain as to what the possible cause might be. Ice skating with my wife's extended family in Tunbridge Wells? There had been a moment when I'd slipped. Possibly one under-knacker had clacked violently against the other, like Newton's Balls, and the resulting force caused some kind of sprain? It was also true that I bicycled a lot to and from work. Presumably that causes some wear and tear on the undercarriage?

The aetiology was obscure but what couldn't be denied was the constant throbbing pain.

I said to my wife Nancy: 'One of my testicles is really swollen. Seriously. Look.'

'Oh Jesus, stop it,' she said.

It was by now a couple of days after Christmas and we had plans to stay with friends in Norfolk. But we agreed I should probably get the testicle checked out while she began the holiday with children. They drove off and I made my way down to an urgent health clinic in west London.

When I was seen, after forty-five minutes or so, the doctor was a young woman, who I had the vague impression might recognise me from television. I sighed inwardly.

I went into her office, or her surgery, or her examining room and, having collected myself, I said: 'Something's up with my testicles. The right one is swollen and painful.'

'OK,' she said, 'we'd better take a look.'

My mind went back to a time years earlier when I'd noticed I had an uncharacteristically itchy bum and I'd gone to the doctor – a man I'd never seen before – and asked him to take a look, and I had the strong impression he thought I was doing it for some weird sexual thrill, especially since he couldn't see anything wrong with my bum.

On this occasion, with the testicle, I thought, at least it's visibly swollen. There is definitely something wrong.

I dropped my trousers, wishing I had worn a smarter pair of boxers. The doctor ran through the usual sorts of questions. Does it feel tender? When did you notice? Then felt it a little bit. 'I'm not too sure,' she said. 'There are two possible things it might be. The thing is, one is pretty straightforward and I can prescribe antibiotics, but the other is more serious and in some cases even fatal. Now, it's not likely to be that but I wouldn't want to take any chances.'

She started making phone calls to various NHS personnel around London, specialising, I guess, in cobblers, and I could hear her being transferred from one department to another. She seemed to be going to a lot of trouble and the thought flashed through my mind that she was aware of the burden of being entrusted with the fate of the UK's premier purveyor of presenter-led documentaries about American subcultures or, more specifically, the fate of his testicle.

This thought undoubtedly says more about my narcissism than anything real.

A little later, I was sent on my way to St Mary's Paddington.

The hospital was busy. There was a wait of several hours. When I was finally seen, I explained the situation, only to be told there had been a mistake: the relevant department was actually at Charing Cross Hospital in Hammersmith. Evidently there had been a mix-up during the transferring of phone calls and I had ended up at the wrong place.

By the time I arrived at Charing Cross Hospital it was dark. I'd spent close to five or six hours waiting and going between medical buildings.

This time I was seen quickly, though. It was a man.

'Do you mind if I get one of my students in here?' he said.

'No, that's fine.'

I dropped my trousers again and the doctor peered at my swollen ball as the student looked on.

'Yep, I've got the picture. Orchitis. Infection of the testicle. Could be an accident. Could even be something you ate. Course of antibiotics should sort that out in a few days.'

Then he added: 'I've got to say, I'm a big fan of your documentaries.'

That evening, I took the train to Norwich, where I joined Nancy and the boys, and the following morning I pushed the little one, Walter, in his pram around a hillside that overlooks the city, conscious of my testicle jostling in my trousers like a spiteful troll. The next night was New Year's Eve. We visited my old friend Adam Buxton and his family at their converted farmhouse, staying up and toasting the year ahead, while I wondered inwardly whether I'd be remortgaging the house and should I just apologise or did that, as the lawyers claimed, lay me open to massive damages.

The next day we drove across to the eastern-most edge of Norfolk, to a little village called Sea Palling, whose buildings were mostly washed away in disastrous floods in 1953 that had killed seven people. Nancy and I and the boys whiled away the hours in an arcade filled with machines that cascaded two-penny pieces and spat out long snakes of tickets that you could trade for prizes and I tried to forget about the legal case.

After a few days of antibiotics, the testicle returned to its accustomed size, presumably a little wistful about its brief visit to the big leagues. And, by a strange quirk of fate, the Miscavige infection went down a few weeks later – finally succumbing to the weeks of high-dosage legal correspondence. Afterwards, along with the relief at the situation having gone away, I had the feeling of having been initiated, and that maybe this was the price of having been credited with more bravery than I deserved. Perhaps, on occasion, you had to weather misfortune that was undeserved – or at least, unglamorous, unexciting and ten times more worrying than an angry glistening wrestler with nipples like rivets or an exasperated Klansman caught out with Nazi figurines.

Other than occasional attempts to hack my emails, which may or may not have emanated from Scientology, or the *News of the World*, or a Russian troll farm, things went largely quiet.

And I've had no further issues with my balls.

MALALA YOUSAFZAI

At a time when I needed them most, the NHS saved my life. The doctors, nurses and staff helped me learn to walk and speak again. They gave me a future, as they are doing for so many others across the UK today.

Every life they save is a gift to our families, and our communities. The NHS is fighting for our future and we have an obligation to make their fight worthwhile – to raise children who care for their neighbours, to fight for truth and save more lives.

Many people have said our world will never be the same after this pandemic. Thanks to the NHS, I hope it will be better.

FRANK SKINNER

I was in Cornwall on a yoga retreat. Yes, I know. It was a passing phase. I pretended it was an attempt to attain some deeper spirituality but, really, I just wanted a flat stomach. Not everyone who does yoga, of course, has such limited aspirations. Some want a nice arse as well. Anyway, my partner, Cath, and I endured the five-hour train journey, mats protruding from our knapsacks, and finally made it to our temporary om from home. We knew most of the people there. We took off our socks and joined in.

On the second night, a Saturday, we decided to hold a mass chanting session at midnight. The start time seemed a tad theatrical to me, but by this stage I had almost completely committed to the vibe. The chanting was to take place in a studio-type building about thirty yards from the house we were sleeping in.

I laid my mat in the studio at about 11.50 p.m. I always found, before any long-duration chanting, I needed a few minutes mumbling as a sort of a ramp. Also, Cath, emanating stress-spores as she frantically searched our room for a scrunchie, was undermining my equilibrium and I needed to get away. I reminded her, with some gravitas, about the significance of the midnight start, even though I wasn't quite sure what that significance actually was. She assured me that, even if she was slightly late, she'd definitely be there for the chorus.

At twelve, the chanting began. I was annoyed that Cath wasn't there and that inner rage gave me a sharp vibrato I could have done without. As the group worked themselves up to a crescendo, I felt less and less part of it. My anger, caused by Cath's absence, had de-yoga-ed me. I stayed to the end but only physically. I was utterly distracted and already working on a few killer lines for the forthcoming argument.

As I left the studio, with my fellow retreaters asking me why Cath hadn't turned up, I almost literally stumbled upon her, lying on the lawn. She was whimpering and saying she'd broken her toe and was in agony. Someone brought out joss sticks and by their dim light we could see that the big toe on Cath's right foot was at a ninety-degree angle to the rest of the foot. It looked like one of those trafficators they used to have on Austin A30s. She explained, through the tears, that she had been racing barefoot across the grass, not wanting to miss the first in-breath and, amid the darkness, had accidentally kicked the handle of an upturned cauldron. She'd lain calling for help but her cries were drowned out by the chanting.

My first thought was, 'If you laugh now, this relationship might end tonight.' My second thought was that I had to get her to an A&E department and I didn't have my car. My third thought was that I wouldn't now be able to use the melodramatic 'I don't want my voice involved in any chant that doesn't involve your voice' line I had been internally preparing.

Getting a taxi in the countryside is like one of those scenes in Hammer horror films where Peter Cushing walks into a village tavern and, unfolding a small sheet of paper from his pocket, asks one of the bar staff if they've ever seen the symbol written on it before. I had to get through a lot of fear, suspicion and downright hostility. I spoke to many people with impenetrable accents who offered information and advice, all of which sounded like extracts from a nineteenth-century ballad. Eventually, after seventy-five minutes, the car arrived.

By the time we reached Truro A&E it was about two o'clock on Sunday morning. I paid the driver the fifty-five quid he assured me was the going rate for that journey. It reminded me of Mr Pickwick buying a horse from a local dealer who received his money 'with a smile that agitated his countenance from one auricular organ to the other'.

Arriving at A&E was like arriving at a nightclub. Young women in almost no clothes and young men whose testosterone was forming beads on their heavily tattooed skin stood outside smoking, snogging and arguing in the light from the hospital windows. I knew if they smelt incense on me I was dead. Still, if I was going to get brutally assaulted anywhere that night, I could do a lot worse than the entrance area of the local A&E.

Cath and I walked through the throng as we had often walked past bulls kept illegally close to public footpaths on our numerous walking holidays. We were relieved to get inside but soon realised we were among the same crowd. They were just resting between cigarettes. The man on the counter told me he loved me, that I was 'the funniest man in Britain' and that we should expect to wait about four-and-a-half hours before Cath could be seen by anyone. I realised I was in a place where celebrity counted for nothing – my own version of Kafkaesque.

There were no seats. I helped Cath to a wall. A woman in a little black dress lay sleeping across three chairs, near to where two men were arguing about something too colloquial for me to understand. Another man was bleeding, quite heavily, from a head wound. I was waiting for Virgil to sidle past with Dante.

Eventually, after what, in fact, turned out to be only three-and-a-half hours, a woman appeared and told us she was the triage. I was very excited by this because I'm always keen to learn new words. I'd never heard of a triage before. She explained her role was evaluating and prioritising the needs of the A&E patients. She smelled of clean linen and competence. I felt Cath relax on my supporting arm. We had found an oasis of calm amid the madness. As she examined Cath's foot, I asked her about the triage business, about A&E, about coping with abuse, drunkenness and vomit on a regular basis. She smiled.

'They're just people,' she said. 'We can usually help them.'

All my fear, my moral outrage, my arrogance, suddenly seemed a bit stupid. She was, as far as I could tell from our ten-minute conversation, someone who wasn't very interested in judging people but was very, very interested in healing them. She didn't even seem to be judging me for judging them.

'So, when is your quietest time?' I asked.

'When it rains,' she said.

I considered the implications of that fact. In any other context, I'd have gone on about how outrageous it was that people were obviously going to A&E with problems that they suddenly considered manageable if the weather became a little overcast. But, by now, such views seemed pathetically inappropriate. I knew I'd be barking up the wrong triage. That smile of hers made me feel slightly ashamed of myself. I didn't want to be Disgusted of Tunbridge Wells any more. I wanted to be all understanding and serene like her. So, in the early hours of Sunday morning at Truro A&E, Cath got her toe strapped up and I got a lesson in compassion, a lesson in acceptance and, ultimately, I suppose, a lesson in humanity. It was the sort of stuff I was supposed to be learning at the yoga retreat. Still, never mind. As long as I got that flat stomach ...

EMMA WATSON

Dear NHS,

I first got to know you when my mum had issues with her diabetic pump when I was a child. My brother and I had to wait on our own while she was being seen. There were toys and it was comfortable and you were kind to us.

I got to know you better when I broke my toe during that dance competition I was training for. You gave me gas and air and helped put it back into a normal (!) looking place. I felt I could trust you.

Later, my brother broke his ribs during a sports match. You took care of him too.

At university I went abroad. My new friend's lung collapsed so I took him to hospital. The doctors wouldn't see him until he could produce insurance papers. All the time he couldn't breathe. I was terrified he'd die while trying to produce the right paperwork – I missed you.

My best friend, her husband, her mum, her dad and her brother all work for you. They are family to me. I want to scream and cry in frustration when I see junior doctors striking because the work and the way the system has been drained makes the job feel impossible and untenable. They are the people I respect the most in my world and I hate that what they want to do, what they love to do, what they feel is honourable to do has been made hard to do or a difficult vocation for them to pursue.

NHS, you've been there for me. You've been there for the people I love. YOU are one of the things that makes me proud to be British and now you are the thing keeping us together during this crisis. How can I possibly thank you enough? Just —

Thank you.

Thank you, thank you, thank you, thank you.

STEPHEN FRY

Unextraordinary

Yes, like any other Briton I have my share of NHS stories. Emergencies, tragi-comedies, disasters and deep trauma feature amongst them. Many will have more dramatic, more eye-watering, more heart-rending, more eye-popping stories to tell than I have. Mine reflect the particular chronic conditions to which I have been subject in mind and body. Nothing too extraordinary. But they are extraordinary *if you transplant them*. I'll come to that.

Story one is ugly and unhappy, but I'll tell it. Aged seventeen, or maybe sixteen and a half, I wake up in a ward in the Norfolk and Norwich Hospital. My stomach has been pumped. I have taken a wild cocktail of pills in a despairing effort to end my life. A psychiatrist comes and holds my hand and talks to me. I am smiled at, comforted and given time to come to myself. There are lots of other details but I won't embarrass myself, my family or you by dredging them up. It's enough to say I was helped and here I still am.

Story two will be familiar to those who, like me, have been laid low by sudden acute episodes of asthma during the course of their lives. They will know that the devil can lash out when you least expect it. For almost all of the year 1988, I house-sat in Islington for the writer Douglas Adams. He was thousands of miles away in the company of the zoologist Mark Carwardine, collecting information on disappearing animal species for their book, *Last Chance to See*.

One evening I invited my friend Ben Elton round for a bite of supper. Sometime after midnight, he reckoned it was time he was going. I picked up the phone and called for a taxi. While we waited for it to arrive, I went to the kitchen to forage for one last drink. I found two bottles of beer. Chimay, the brand was called, brewed by Trappist monks in Belgium. Very fancy. I poured out two glasses. Cloudy. As soon as the fluid touched my throat I recognised the signs. By the time I had put the glass down I could barely breathe. That's how quick the reaction can be. My lungs just seemed to completely close down.

In shock, Ben tried to call 999. He didn't realise that Douglas's up-to-the-minute house telephone system required the input of an initial 9 for an outside line. So Ben needed to dial 9-999. I didn't have the breath to tell him. Or to do the dialling myself. Fortunately, just at that moment, Ben's cab pulled up outside. I dimly remember being dragged along to the front door and pushed into the taxi and the sound of Ben's urgent voice, rising in panic: 'The nearest hospital! It's an emergency!'

The next thing I recall was waking up in a recovery room, oxygen mask on my face, concerned eyes peering down. It had been a close-run thing, it seems. Not much oxygen had been getting through to my brain by the time the taxi had arrived with a squeal of brakes at the main entrance of St Bartholomew's Hospital. Ben had got me onto a wheelchair and run in, yelling for assistance. By the middle of the afternoon of the following day, I was all better and was released with a brand new EpiPen and a snazzy portable nebuliser in case it should happen again. Follow-ups included visits to all kinds of specialists, right up to Professor Malcolm Green, now retired but then head of the National Heart and Lung Institute.

I have, as you probably do, a good many stories like that. Broken ribs, arms, strange eruptions of one kind or another. Spots in front of the eyes, a drop in hearing, a lump here, a limp there. Nothing, as I say, extraordinary, but certainly extraordinary *when you transplant them*.

When you transplant them, for example, to the USA. I live part of the year in California. Tell American friends how you can arrive at a surgery or hospital in need of attention and walk out having received kindly and professional care without at any point being asked for a credit card number. Explain that at no point are you handed a bill. That at no point do you frantically have to get in touch with an insurance company. Even if, as I always manage to do, you forget your national insurance or NHS number, you are calmly assured that the paperwork will catch up.

On one occasion in America, I was so incensed by the die-hard bloviated twaddle being talked on right-wing radio broadcaster Rush Limbaugh's programme that I called it up. He and his loyal listeners had been railing against the threat of Obama's 'socialised medicine'. How ineffective and iniquitous and foolish it was. What an affront to the honest taxpayer. How in Britain people were always complaining about it. The waiting lists. The sick lying on gurneys in corridors. How it made no sense to put your life in the hands of *the government*.

I phoned in and told Limbaugh that I reckoned if you stopped Britons on the street it would take you a long time to find one who didn't love, honour, venerate and feel deep pride in our National Health Service. 'In America you have a "socialised" military after all,' I said. 'Taxpayer funded and defending America against the threats of enemies foreign and domestic. Well, we have a medical army defending us against the threat of sickness and disease. Why is one acceptable and the other not?' Limbaugh said something like, 'You're very confused, my friend, very confused indeed . . .' before I was cut off. The 'my friend' rankled especially.

Well, how other countries choose to look after their sick (or not to) is none of my business. But we in Britain need to remind ourselves from time to time how our individually unextraordinary stories add up to something so highly extraordinary after all. Did we need a pandemic to be reminded?

SIR MICHAEL PALIN

My experiences of the NHS have the common theme of humanity and humour. At the tender age of twenty-four, I had my appendix removed at University College Hospital in London, which subsequently won a special place in our hearts as the birthplace of all our three children.

It was my first experience of life on a ward. I was writing comedy at the time but my three-day stay was a sitcom in itself. The patient next to me was a quiet and courteous man, but the nights were difficult for him. As soon as it was dark and silent he'd cry out in a querulous voice, 'I'm dying!' If there was no answer, and there never was first time, he'd repeat it more powerfully.

'Nurse! I'm dying!'

'No you're not, Mr Lygoe,' the nurse would reply patiently.

'I am,' Mr Lygoe insisted.

'No, you're not.'

'Yes, I am!'

This would go on until the whole ward was awake, including the man in a bed opposite, who had two very contradictory problems. One was a terrific sense of humour and the other was recently completed surgery on his haemorrhoids. One thing he shouldn't be doing was laughing and the one thing he wanted to do more than anything else was to laugh. And not just a snigger or a giggle, these were big full-throated belly laughs, each one accompanied with the agonising cry. 'Oww … Oh my God!' which would make him laugh even more.

He found everything hilarious. One morning my friend Barry arrived with a plastic bag containing *Playboy* magazines going back three years.

'Don't open those, Michael,' he called across the ward. 'You'll split your stitches!' This set him off again. 'Haa! Ha! Oww! Oh *my God*!' I turned the pages, licking my lips as I did so, which was cruel.

A chaplain came to visit me. Nice and well-meaning but no match for the priest who bent low over his Catholic patients, offering copious words of comfort, never loud enough to completely mask the sound of Guinness bottles being stowed in the bedside cupboard.

At a table in the middle of the ward, a burly man in a white coat was using a hole-puncher to prepare papers for a file. He was appearing to make what seemed a simple process into an Herculean task, raising his arm high in the air and bringing it down with a force that shook the table.

I asked a passing nurse who he was.

'Oh, him?' she replied brightly. 'He's the surgeon who took your appendix out.'

My two other experiences both involved my good friend Terry Jones. He was preparing some oysters for one of the wonderful meals he used to turn out, when the knife slipped and dug deep into his hand. We wrapped it up as best we could and I drove him round to King's College Hospital in my Mini, with Terry's arm held up vertically through the sun roof.

A&E at King's was a pretty grim place that day, with all sorts of people in various states of distress, but when they saw two Pythons coming in, one with a bloodied arm sticking straight up in the air, it was as if Christmas had come, and appreciative laughter rolled around the room.

It's good to be able to raise others' spirits. When I was in Bart's recently, recovering from heart surgery, a regular stream of doctors kept coming in to check my stats. I couldn't believe the attention I was getting. After an hour or so, the staff nurse came in and apologised profusely for the presence of so many doctors.

'We've put a sign on the door now,' he assured me. 'You won't get any more.' But didn't putting a sign up to keep doctors out seem a little counter-productive? 'Oh, they're not your doctors,' he explained, 'they're just fans.'

A final tribute to the lengths NHS surgeons will go to to accommodate patients' wishes was at a hospital that had better remain nameless. Terry Jones was having some of his bowel removed and expressed a wish that at some point during the procedure they might take a photo of his bowels. He showed me with enormous pleasure a picture of a little coil of bowel posed neatly on his tummy. Terry thought this was marvellous.

'You so rarely see your own bowels.'

He showed it to everybody and even considered using it as his new fan photo.

The NHS is basically made up of people looking after other people. Though there is much skill and technical support required in treatment, the most important thing is to see patients not as numbers on a screen but as fellow humans with all the idiosyncrasies that make each one of us different and special.

In my experience, they do this brilliantly.

DAVID TENNANT

27 November 2019

It was my first day on a new job, filming a drama for ITV in London. It had been a long time in the planning and development. I'd been talking to the director about it for around four years and we were finally on set. It was based on a true story; many books had been read, witnesses interviewed, documentaries consumed. I was all set to go deep for a few intense weeks of filming. Long days at work all wrapped up in my own, ever so slightly indulgent, process. We were about an hour in when my wife called.

'Nothing to worry about but I'm just running the baby to the hospital. She had a bit of a temperature and the GP has *insisted* we get her checked out.'

I could hear the sigh of inconvenience in Georgia's voice at this jobsworth doctor forcing her to take time out of her hectic-enough life to go down to Chelsea and Westminster on a cold Wednesday evening. This was our fifth child; we knew an overreaction when we came across one. I went back to work and didn't feel particularly concerned.

Through the evening, the calls got a little more alarming. I kept filming as best I could, but as the night wore on I got more and more distracted, as our seven-week-old baby did not get quickly discharged from hospital.

Emergency childcare was hastily called upon. Who could cover the school run? Is the baby staying in overnight? What's going on? When I was done at around midnight I went straight to the hospital and suddenly the new job felt a lot less important.

The week that followed is a bit of a blur. I would leave hospital to go filming but all my good intentions about being deep in character evaporated in a fuzz of real life. I just had to get through each day to get back to the hospital. The drama wasn't pretend any more.

When it was all over my wife wrote this on social media:

> Last Wednesday my NHS GP sent my baby to A&E after spiking a high temperature. Eyes rolling at this seemingly over-the-top reaction, reluctantly I trudged along, mildly coughing child in tow, to my nearest NHS hospital. We were seen within 10 minutes. After being checked over by 2 'over the top' NHS nurses and another 2 'over the top' NHS doctors, the now slightly lethargic baby was admitted. What ensued over the subsequent 6 days will haunt me forever but now back home, on the sofa, my baby tube-free and pink again I take away one thing; our

> NHS is magic. An underfunded, understaffed and under threat sort of magic. Full of amazing people whose 'over the topness' puts people back on sofas together. I can't thank you enough NHS and from now on my family will do all it can to help keep you together. Just as you did for us.

To live in a country where we can take remarkable, life-saving care so utterly for granted that we can find being *so* looked after almost inconvenient is a luxury and a privilege I will forever be grateful for. Our baby is fine now – she won't remember anything of her week in high dependency. But we will never forget it.

DAVID BADDIEL

It was Christmas Day 2015 when my partner cut her finger off. She didn't do this festively. She didn't do it peeling sprouts or slicing turkey: all that (plus a nut roast for her and my vegan daughter) had already been lovingly prepared.

We were literally about to sit down to Christmas dinner when the finger came off. This was in Cornwall, where my partner Morwenna Banks – writer, actor and Mummy Pig in *Peppa Pig* (always complicated for a Jewish bloke) – is from. We were staying in an old house by the sea. It was, as the maritime stories say, a wild and stormy night.

Outside that property is a shed, which operates as a utility room. Morwenna, always keen to utilise spare time, had thought, 'Dinner's nearly ready; I'll just go and get the clothes out of the dryer.' But on coming out, as she was trying to close the ancient and clunky shed door while holding a basket of clothes, the wind took the door and slammed it shut on her left hand. She came back into the house and very calmly told me that the top of her left index finger had been taken off.

At first, even though there seemed to be a *lot* of blood in the sink after she washed it off, her calmness made me think she meant perhaps a tiny sliver of skin. However, when I went out to examine the scene of the accident, I saw on the ground outside the shed door what unquestionably was a proper, joke-shop section of finger. Swinging the torchlight onto it was as close as I've come to being in a horror film. Until coronavirus transformed all our lives into something a bit like *Dawn of the Dead*, of course.

Back in the house, I did the obviously right thing: I put the bit of finger on some ice in a Jiffy bag and left it in the freezer (on top, I think, of an inordinately large tub of Roskilly's vanilla ice cream). That, I knew, would mean that, should it be needed in the future, the fingertip could be sewn back on, no bother. We called NHS Direct. The woman at the other end of the line was worried. She said we should go to hospital.

But Morwenna wasn't keen. Morwenna is a selfless person and absurdly stoic. And one thing about Christmas 2015 was that my mum had died on 20 December 2014, which had cast a bit of a shadow over our festivities the previous year. Plus Morwenna's mum had died very soon after, in January. So there was a lot of pressure to make Christmas 2015 especially brilliant for our still-young children. Perhaps I can make this clear through our then eleven-year-old son's tearful response to the door accident:

'I just want to have a Christmas where no one's finger gets cut off and no one dies!' Which frankly was closing the stable door after the horse had bolted. And probably, in so doing, cutting off the stable-owner's finger.

This was why she very much wanted *not* to spoil Christmas for the kids this year. Plus, I, who am not selfless, had cooked the turkey and was really looking forward to eating it (mainly brown meat for me). So we bandaged up her hand as best we could and had Christmas dinner. Towards the end of which, it became clear to me that Morwenna couldn't pass the brandy butter without fainting.

We called NHS Direct again. The same woman told us now that because the door was old and rusty, and because Morwenna was starting to feel woozy, we HAD to go to hospital, as there was a danger of sepsis. Which meant that, after a certain amount of apologising to the kids for another ruined Christmas, we got in the car.

When I say the car, what I mean is MUD. We go to Cornwall a lot, and earlier that year I had invested in a car to keep down there, in Morwenna's cousin's garage in Falmouth. The term 'invested' is being used very loosely here. With two kids and large distances across a county to travel, I should of course have bought a second-hand Ford Focus or Clio or something else reasonably reliable. Instead I bought a fifteen-year-old Volkswagen Golf convertible. In my mind this was a classic car. It wasn't. It was in fact neither a classic car, nor a reliable second-hand runaround.

It was a piece of shit. It cost me a thousand quid. It often – very often – did not work at all.

What was particularly crap about MUD (so called because those were fittingly the first three letters of its registration) was its canvas roof. It leaked, badly. I had dealt with this not by getting it fixed in a garage but by taping a shower curtain under the roof in the places where I thought the leaks were. Which worked brilliantly. Except when it rained. Which in Cornwall, on Christmas night 2015, it was very much doing.

We got in the car – me, Morwenna and the children, Ezra and Dolly. I heard Morwenna, through her pain and wooziness, say: 'My seat is really wet.' I made a mental note to perhaps tape *two* shower curtains to the roof in future. I turned the key and, amazingly, MUD did at least start.

We were staying quite far south on the Lizard peninsula. We had discussed which hospital was best. Treliske in Truro was the biggest but it wasn't that near, compared to smaller ones in Helston and Penzance. We headed first therefore to Helston, where there is a Minor Injuries Unit; we decided, optimistically, that this was what had happened to Morwenna's hand. When we got there it was one minute to eight in the evening.

We rushed in. It was Christmassy. There were fairy lights. A woman on reception was dressed as an elf. Another woman, a nurse, had plastic antlers on. However, the Helston Minor Injuries Unit shuts at 8 p.m. (this, you see, is why I remembered the very specific time in the previous paragraph) and they were in the process of doing exactly that: shutting. They said: it's not a minor injury. Go to Truro.

So obviously we went to Penzance. This is because, by now, the wind was really bad and the rain was coming down heavily, and Penzance A&E was a bit nearer than Truro. The roof was leaking so badly by now that the whole family was sodden: it was like I had chosen to drive on through a storm with the convertible roof down.

When we got there the hospital was, at least, open. I proudly showed a doctor who came to examine Morwenna the fingertip I had remembered to bring, still in its Jiffy bag. He looked at it and said, yeah, you may as well throw that away. Apparently, the best thing to do with a bit of your body that's been cut off is to keep it warm. Icing it up kills all the nerves. I think he actually threw it in a bin in front of me.

Morwenna was taken to a cubicle and given gas and air. She told me afterwards that while she was in there waiting, she heard the duty doctor say to another woman, behind the next curtain, 'So … is this a regular problem for you? Cystitis?' I stayed out in reception with our disconsolate children. Eventually, she came back and said: we have to go to Truro.

They had cleaned and properly bandaged the finger. But it turned out she had broken the bone in the top of her finger and Penzance Hospital did not fix broken bones.

Just to be clear, for those of you who don't know the geography of Cornwall, Truro is in the *opposite* direction from Helston to Penzance. We had driven about an hour in the opposite direction of where we actually needed to be. So we then had to drive two hours back across the county to get to Treliske Hospital in Truro, where we could have gone in the first place. I began to wonder, truly, if the shower curtain was going to survive.

By the time we got to Treliske, it was about 10 p.m. We sat in A&E for some time. The thing I remember most clearly is that they had a big telly in A&E up on the wall and *Michael McIntyre's Christmas Comedy Roadshow* was on. I like Michael. But I have never hated his jolly, smiling king of light entertainment face more. After another hour, Morwenna was taken away; this time, it transpired, for the night. Because here, finally, the doctors could fix her finger, or at least, they could, as they put it, 'nibble the bone' – at which point I thought *I* was going to faint – down so that it might one day work as a finger again. But it would require her staying in for the night and not coming out until Boxing Day evening.

Me and the children said a tearful goodbye to Morwenna, got back in MUD, and drove *another* hour back to our house in the Lizard. I'm an atheist, and Jewish, and I spent most of it praying to Baby Jesus that the car, whose engine was now sounding worse than someone with a COVID-19 dry cough, make it back. We did, but by the time we arrived, it was well past midnight, which meant that Christmas Day had ended for my children somewhere on a rain-lashed A30 in a shit car under a shower curtain that was now more of a shower than a curtain.

I'll be honest. I'm not sure it was a better Christmas than the one when my mum died.

So. Despite it being Christmas, this isn't a story about miracles. Nor about the NHS being miraculous. They didn't keep the Minor Injuries Unit open for us at Helston. They couldn't fix Morwenna's finger in Penzance. In Truro A&E they wouldn't even switch off *Michael McIntyre's Christmas Comedy Roadshow*, no matter how many times I asked.

But those women dressed as elves and reindeers in Helston had been working all day – all Christmas Day. The doctor in Penzance was patient and kindly spoken, and betrayed almost no sign of incredulity at having to deal with a woman who had come into A&E on Christmas Day with cystitis. At Truro, they fixed Morwenna's finger and by Boxing Day she was out, with a huge bandage all the way up her arm, and Christmas continued. Her hand, in the end, was fine. And perhaps most importantly, that woman, whose name I will never know, but who, again on Christmas fucking Day, was working the line on NHS Direct and who told us to go to hospital or else Morwenna might get sepsis – she may well have saved my partner's life.

And frankly, if Morwenna *had* died, it really would've been a shit Christmas.

CAITLIN MORAN

All church halls smell the same – a combination of hot dust, floor polish and tea. All church halls sound the same,whether they're holding a jumble sale or collecting votes on polling day: a polite British murmur, occasionally punctured by a baby yelling and the noise of the pensionable samovar clanking into action.

But today – today, the church hall *looks* different. Today is an event that, in this place, comes around three times a year – because it's Blood Donation Day. And so there are signs in the street that read 'DONATE BLOOD HERE TODAY' and two huge blood donation vans in the car park. And, when you enter the hall, there are nurses in blue tabards servicing the twelve reclining chairs where people lie giving blood; as those who wait murmur, a baby occasionally yells and the pensionable samovar clanks into action.

Outside, it's beautiful, a beautiful day – shoals of swifts screaming as they chase insects; horse chestnuts in full bloom. But in here, people have cheerfully given up their picnics, their walks, their lunch hours and their housework to give a pint of blood. To just fill a 500ml, clear plastic bag, which is then carefully labelled and stored in refrigerated units until needed.

One man has brought his toddler – with one hand he keeps a bottle jammed in her mouth, while the other arm is tethered by a needle. Another man, in a suit, turns up. He looks like he spends most of his day shouting at people on the phone, making big things happen. He has an air of peevishness – as if, in his time, he has sent back a *lot* of soup.

Today, however, he sits patiently on a chair, waiting. He is greeted as a regular. Here, he's taking a holiday from being a bit of a bastard. It looks like a relief to him. As if perhaps most of the time he has *too much* blood, from all the red wine and steak, and it's quite calming to have it drained away and put to use inside someone less ... *driven*.

I haven't given blood for – God – eighteen years? First there were babies, then work and then, on three pivotal occasions, I was drunk. I feel ashamed I have left it this long. Giving blood is something I *believe* in – like I believe in libraries, and council housing, and David Bowie. So I guess I'd become a ... *lapsed* blood donor? But today, I am coming back into the faith.

I know why I want to give blood – for donation feels like an act of thankfulness. It acknowledges that you are alive, and grateful for it, and wish to share the gift of living with someone else for whom living has become, suddenly, perilous. It is the most useful thing you can do in just twenty minutes. It is *fascinating* to be this useful. To see how something you can spare – a pint of blood – is so treasured that this whole room has assembled to take it from you.

I go into a booth with a nurse and she does a finger prick – dropping a bead of blood into copper sulphate.

'Iron test,' she says. The bead sinks. My blood is heavy with iron. How amazing to be full of metal. To see it.

On the reclining chair, the needle is bigger than I remember it – hollow, like a tunnel being pushed into a ring-road of arteries and veins. I make myself watch, because this is interesting, too. It hurts as much as pulling out a single eyebrow hair. That's the exact amount of hurt. I look around at everyone – all calmly lying back as they donate, as Tony Hancock put it, nearly an armful.

I'm very surprised when I start to cry. It's one of those cries that just enters with no warning – like when Rik Mayall makes a cameo in *Blackadder*.

It's because, in the last few years, we have been led to believe that we are a little bit harried, a little bit unyielding – that the world is going to hell in a handbasket, and the smart thing to do is harden your heart and look out for yourself. We've split into camps, tribes – to begin to talk is to fall into an argument and reveal yourself as someone else's problem, or enemy. Baby boomers are pitted against millennials, Leavers against Remainers – and yet, even in our furiously bonded groups, we've never felt more anxious or alone. Along with sparrows, bees and skylarks, it feels as if love is in decline too. You do not see it around so much any more. You do not open the door and hear it singing.

But here, this room is full of the least talked-about love – love for someone you've never met. Here is a system set up, without profit or material reward, based on a simple idea of a country never wanting to see someone bleed out on a table when there were a thousand people out there who would have given their blood in a literal heartbeat, if they'd been asked.

This, then, is where you are asked. This is where you can lie on the bed and scrunch your hand into a fist, over and over, sending all the luck in the world to the team who will, one day – one terrible, unlucky, critical day for someone – break open the seal on your bag and try to keep someone alive. Maybe *this* heartbeat will turn someone's lips from blue to pink again. Or *this* beat bring back a mother, or a father, or a child. All the calmness and love in this room is being sent into some furious, terrified future in A&E that you will never know about – but where you will be the magic that stops a life from being undone. Perhaps the life of someone you know. Perhaps your own. Perhaps you are not donating at all – but lending. As others once, maybe, lent to you.

The church hall, the vans, the nurses, the donors, the samovar, clunking; Britain thinks it is having an identity crisis but a country is, simply, what is does – and it does this. It has an NHS where people donate their blood for free. Where they come and put 500ml of love in a clear plastic bag, which is carefully labelled and stored against future terror.

In church halls, like this, in every county, smelling of hot dust and love.

DAWN FRENCH

Mum

I hate Derriford Hospital.

It has been the setting for so many life-changing and life-ending moments in my family.

Every time I enter the huge sliding doors at the entrance, my stomach lurches.

It's the muscle memory of all the nerve-wracking, heart-cracking times I've been there.

My nephew and niece were born there.

My old granny Lil was nursed there.

My father-in-law had huge surgery there.

I have visited cousins and uncles and friends there.

So many lives I value have been saved there.

But each time I set foot in there, the most potent memory is of my beautiful mum's beautiful death.

It is possible to do death right.

I know.

I saw it with my own eyes.

At Derriford Hospital.

Eight years ago.

Mum was short of breath.

I think she knew.

She didn't tell us for weeks.

She'd been a smoker since she was thirteen years old.

Loved it.

The kind cancer doc carefully explained that her lungs were struggling.

They offered chemo.

But before even the first appointment, she went downhill rapidly.

They explained it was terminal.

And probably soon.

In the car, she said, 'It's win win, either I stay here a bit longer with all of you. Or. I go and see your dad. At last.'

He'd died thirty-five years before.

She'd missed him so much.

For so long.

Her faith was unwavering.

She would be reunited with him.

A week or so later, on Mother's Day, she woke up and could hardly breathe.

The ambulance came.

She looked out at her beloved sea view, and we all knew she was bidding it farewell.

She went into Derriford.

Brent Ward.

The names of the wards are strangely comforting.

They are places we've been, we know well: Stonehouse, Postbridge, Burrator.

Mum tells us she's not afraid to die, but she's terrified at the prospect of not being able to breathe.

A difficult night, while she struggles with it all, despite lots of loving support.

By the morning, her mind is made up.

Her time has come.

She asks to see her grandchildren.

She tells them all how unique and marvellous they are.

She sees Dr Mary Nugent, who is in charge of her palliative care.

Dr Nugent says 'I know who you are, Roma French, you have done so much for people in this city. It's my honour to help you now.'

Mum says, 'And I know you, Mary Nugent, you are a superb doctor, and I want to go to sleep and not wake up. Please. Thank you.'

Dr Nugent explains the Liverpool Care Pathway.

Mum says that is PERFECT.

My brother and I tell her that we love her very much and that she has been the best mum ever and that we owe our happiness to her.

She smiles.

She takes off her wedding ring.

She gives it to me, and winks.

Her comfort medication is given carefully, slowly and she sinks into a deep deep sleep.

My brother and I take it in turns to sit with her.

Two days.

Three nights.

The nurses let us snatch some sleep on a small bed in the back room when we can.

I am fast asleep when my brother comes in to gently wake me in the early hours.

'She's gone, Moo.'

We stand together next to her for a while.

Soft, grateful tears.

Mum has taught us that it's OK to die.

So many nurses, doctors, cleaners and helpers had been there to support her, and us.

Quietly, respectfully doing their jobs so well, and helping us all to pass through this difficult, sacred, unforgettable moment the best they could.

My brother and I left as the sun rose.

We both took deep lungfuls of cold air as we walked back out of the huge sliding doors.

Plymouth air.

Which Mum had longed to gulp.

The air of home.

Yep, I hate Derriford Hospital.

But try telling me there's a better hospital for my family.

You can't.

Because there isn't.

It's sort of miraculous there.

PAM AYRES

2020

I've been to other countries where you pray you don't get sick,
Where if you're taken ill, no kindly ambulance comes quick,
No motorbiking paramedic roaring through the rubble,
Where if you have no cash to pay then mister, you're in trouble.

We have a gentler system, which is comforting to all,
It strives for our wellbeing, be we elderly or small.
With expertise, professional, extended countrywide,
So, in an emergency, a world is at our side:

Consultants and anaesthetists and those who man the door,
Specialists and surgeons and the folk who mop the floor,
The porters and the nursing staff who labour night and day,
And never ask the patient if they have the means to pay.

A plague has come, a plague that's new, yet old, as old as time,
Fomented out of suffering, and cruelty, and grime,
With unimagined images which linger in the head,
Refrigerated trailers for the storage of the dead.

With calm and regulated care, staff with one accord,
Though fearing for their families, are working on the ward,
Where end-of-life care nurses, in their strange protective dress,
Hold a fading hand to dull the pain of loneliness.

Thanks to every doctor, every midwife, every nurse,
Every single worker in the fight for life immersed
Whatever God you recognise, may your endeavours bless,
Sending love and gratitude to you. The NHS.

JOANNA LUMLEY

I am in an awful hole here. Being as fit as a flea and virtually unsinkable, I have no NHS experience AT ALL. (Well, I do; when I had my appendix taken out when I was twelve and was put on an old people's ward, but that was sixty years ago and times and the NHS have changed mightily.)

So, I haven't got a story to tell but I LOVE the NHS with all my heart, and love paying my taxes in full because I know it will benefit. I am tear-streaked not to be able to send more than a HUGE virtual hug for the whole of the NHS, and the deepest curtsey, and the most colossal gratitude and admiration.

In fact, if I heaped up all my admiration and affection for the NHS and pushed it towards this fantastic organisation's feet it would simply disappear from view.

JO BRAND

Hello everyone, I'm Jo Brand and I worked as a mental health nurse between 1978 and 1988. I loved it: the camaraderie, the closeness of our team working together, the laughs, the adrenalin, the successes, the friendships, the canteen sandwiches, the cigarettes, Christmas and even the bad bits. And to anyone working in the NHS now, I salute you.

Here are a few stories from my time as a nurse ... Yes, it was back in the olden days, even before mobile phones.

EMERGENCY

As a student mental health nurse, I inevitably got my turn on the emergency bleep a few months into my training. It bleeps, you call a number, you run to the ward where you're needed. I turned up at one emergency to find a recently arrived police cadet holding down a protesting, wild-eyed woman in her sixties. Having got to the ward and witnessed this woman shouting and waving her arms around, he immediately and efficiently got her down on the floor and lay across her.

All great, except this woman was the senior sister.

UNEXPECTED

After a gruelling couple of hours on the Emergency Clinic, where we tried to persuade a very ill, threatening, paranoid young man to come into hospital, and in the end were forced to physically restrain him in the Outpatients while people waited for their appointments, I was glad to get him onto a ward safely. The whole incident had been disturbing and upsetting for everyone. About a month later, the young guy in question was unrecognisable when he arrived in the clinic to thank us for looking after him and shook everyone's hand. This is the reason most people become a mental health nurse.

TEACHING SESSION

As student nurses we didn't think of ourselves as daunting. Until one day, a junior doctor came to give us a lecture in the School of Nursing using a teaching aid which involved writing on an acetate sheet, which is then projected onto the wall. He switched on the teaching aid and, as we waited for him to write, he walked towards the rectangle of light on the wall and started to write on that. Poor sod. I think it took him a long time to live that down.

UNDER LOCK AND KEY

On secondment as a student to a mental health unit for the elderly, I was asked to take a woman in her seventies, who I didn't know very well, home for the day in order to assess her ability to manage domestic tasks and see how she coped generally in the home environment.

Please dump all your preconceptions about the frailness of the elderly. This woman, let's call her Enid, was built like a weightlifter with the strength to match. And I wasn't exactly willowy. We were dropped off at 9 a.m. by the lovely minibus driver who was due back at 3 p.m. and then began the strangest of days.

After we got into her ground-floor flat, she didn't even take off her coat before she started pushing the furniture up against the door. Paranoia was a feature of her illness. My enquiry as to why she had done this caused a furious reaction and she stopped for a second to punch me. Quite hard. It hurt a lot. She was bigger than me, tougher than me and scary. I realised I wasn't going to be able to stop her without getting a fair bit of grief, so I decided to let her carry on. I couldn't leave so I just had to make the best of it.

I sat around in this atmosphere of threat for the next six hours, trying to keep it light and chatty with varying degrees of success. The minibus driver arrived back at 3 p.m. and, when we didn't appear, came to investigate. The door was barricaded but he managed to communicate through a window while Enid smacked me round the back of the head. While I distracted Enid, he managed to force the window a bit more and get in. He knew Enid well and she was only too happy to go with him.

NURSES' HOME

One night, I was in my room at the nurses' home and, I have to say, quite drunk after a session in the pub. I couldn't sleep and I'm not sure why but I became convinced there was someone outside my door. Unable to ignore the very strong feeling but thinking I was just imagining it, I opened my door. A man's face was about an inch from mine. Struck with fear, I immediately slammed the door and pushed my bed up against it.

I sat up all night too scared to go out of the room.

In the morning he had gone.

A RAY OF LIGHT

The locked ward could be a harsh place. Most people didn't want to be there and weren't allowed to leave. In-patients were expected to attend a group every morning and, as most people were so ill, this could be frustrating and occasionally scary. One morning, into the midst of this chaos walked a well-known actor, brought in the night before by the police. Still in a hospital-issue dressing gown, he climbed onto a table – our hearts missed a beat – but he began to sing 'Oh, What a Beautiful Morning', from some musical or other, while conducting with an imaginary baton. Some were transfixed, some indifferent, but most joined in and it was one of the loveliest things I've ever seen.

JONATHAN ROSS

I was born in 1960 so have never known what life might be like without the magnificent safety net that the NHS provides. My time on the planet would certainly have been different – indeed, I would have been different. Chances are, I would be missing the top of the index finger on my right hand.

When I was about eighteen months old, my mother left me in the care of my elder brothers, both being more than capable of looking out for a toddler, as they were nearly four and five respectively. But needs must; we had no food, so my mum went on one of her fairly regular foraging trips to the neighbours asking to borrow some. My memory is, of course, not to be relied on, but I imagine hunger got the better of me which is why I apparently crawled to the bin in the kitchen and dug out an old, opened can of beans. I sliced the tip of my finger off and a doctor, whom I'm sure never received sufficient thanks, sewed it back on. Sewing the tip of a screaming, wriggling baby's finger back on can't be easy but he did a pretty good job. It's wonky, sure, but it works. So belatedly, thank you. My career as a hand model never happened of course, but the surgery was free and the finger is functional and that's all I really care about.

Without the National Health Service I might also have an enormous growth on my forehead. Fast forward about fourteen years and you will find me sitting on a bus – the 262 I think – heading to my local hospital, Whipps Cross in East London. You will notice I am clutching a milk bottle on my lap and am getting slightly strange looks from my fellow passengers. The milk bottle, capped off with a rudimentary lid of kitchen foil, contains a pint of my still-warm piss. My doctor, Mr Patel, had sent me in after failing to work out what the growth on my forehead was. But he was confident it should be removed.

You will have surmised by now that my parents, though loving, had a casual approach to childcare and I was effectively self-raised. So being told to bring a urine sample on my visit, I had taken the initiative and filled the whole bottle. Hindsight is a wonderful thing, and carrying a small pot would have been much easier and much less splashy. The nurse who took the pint off my hands presumably absorbed enough information from my look – ill-fitting hand-me-downs, worn-out shoes and a garnish of piss splashed in my lap (that foil lid was a bad idea) – to have worked out I was not a rich tourist slumming it. She reassured me that a whole pint was perfect and would I take a seat.

The lump on my head turned out to be a benign cyst and was removed about a month later. I spent the night on the ward – the only time in my life I've been hospitalised, but I have only the fondest memory of it. The food was good, the people who worked there were universally lovely and my fellow inmates – patients? – who were also having cysts removed were good-natured and funny. I think we all knew how fortunate we were to get this kind of care without needing to pay or get into debt. I woke up with my head bandaged which made me feel roguish and piratical and also lucky – more so when I noticed the two elderly men in the beds that flanked mine both had heavily bandaged groins.

But the point, the simple point, is that someone like me from a piss-poor background – although literally I suppose I was piss-rich, seeing as I could splash free pints around, so cash-poor – could have such complicated and expensive procedures carried out for free. Without wishing to pile it on too much, it's not an overstatement to say that I knew that, even if my parents weren't totally there for me, the NHS always was and always would be.

I am lucky enough to have enjoyed good health most of my life, so mainly I have entered A&E due to stupid accidents that were a) entirely my own fault and b) could have been easily avoided. Case in point: a visit made when I had twisted my ankle painfully and needed relief. I rushed down with my wife and our baby, carried safely in her arms. When the junior doctor asked how I had come to be injured, I unwisely told the truth. We had been watching *Stars in Their Eyes* and I had been so incensed by one of the participants' poor attempt to capture the idiosyncratic dance moves of Mick Hucknall that I had leaped up to demonstrate how it SHOULD be done and discovered, painfully, that it was harder than I thought. After an x-ray showed that it was merely a sprain, I was bandaged and sent home. They gave me painkillers and the staff decided that, due to my bravery, I also deserved a lollipop and a certificate normally reserved for far younger patients.

There's something quintessentially and wonderfully British about both the NHS and our ability to take the piss in a kind way. So thank you all.

KEVIN BRIDGES

I've been asked to help out in a lot of ways since the Covid-19 lockdown began and yet, aside from a few poor attempts at doing keepy-ups with a toilet roll as part of the 'show off your massive garden' series of celebrity challenges, I've found myself gradually descending into a pit of unmotivated slothfulness. Adam has only asked me to contribute a chapter, or even a few paragraphs, with a fairly generous two-week deadline and even at that I'm cutting it slightly fine.

My circadian rhythm is bordering on the alarming, even by a stand-up comedian's standards. I'm not sure if my mind is officially in British Summer Time; I'll need to take a look at a map later and determine approximately which time zone I'm living in. Somewhere near the east coast of America, I'd say. Greenland or somewhere like that. Of course, this is the first time I've opened up about this unproductive rut; my WhatsApp group chats are the first to know of anything meaningful I've accomplished during this bleak period – screenshots of my mediocre times on the few 5k runs I've attempted, photos of some home gym circuits and the front covers of a few books I've bought but haven't started, complete with a bookmark strategically placed to imply progress.

Despite the vicissitudes that the world is experiencing, I'm attempting to maintain some equanimity by focusing on little tasks, no matter how picayune. Like writing down words I've read and didn't know their definitions, and then learning them by looking them up and using them in a sentence. Words like vicissitudes, equanimity and picayune.

Like most people, I look forward to relative normality being restored and I know how important it is that we all do our job, or 'our bit', by staying indoors – no matter how mentally challenging it can be at times – and by checking in, via phone and video calls, on our family, friends and neighbours, especially those who are elderly and vulnerable in any way. And of course, if you're in a position to, make sure you drop off food and medicines and help out any charitable causes which will be struggling massively.

So, to the National Health Service. I would have written 'NHS' but I'm attempting to bump up the word count of my ramblings here – if I hit a thousand words, I can justify an afternoon bottle of some mental-sounding and fairly potent Belgian beer. My wife, Kerry, and I have been out in the garden for the last few Thursdays, applauding and drinking one of said mental-sounding and fairly potent

Belgian beers and prosecco, respectively, in your honour. Like the great people we are. We also pay our taxes in the hope you are provided with the wages, equipment and environments you deserve, which is a more tangible way of showing appreciation, but that's a rant for another time.

Aside from an asthma attack when I was seven, when I was rushed to hospital and kept in for a few days, and a fairly embarrassing operation a year ago (I'll omit the full details here but google 'torn fissure nerve' for an idea. Never strain too hard during a bowel movement, folks), my own personal experiences with the NHS are fairly limited, thankfully. A situation that will gradually change as I grow older and my west of Scotland upbringing takes its toll.

But I write a personal thank you on behalf of some very special people in my life. Notably my parents and my mother-in-law, Joanne, who has been relentlessly and heroically fighting a battle with cancer for many years.

As these are grim times, I feel a more light-hearted and recent NHS anecdote is a better choice. An afternoon last summer, I'd popped in to see my mum and dad, Patricia and Andy; a fairly routine visit, but I wisely came armed with updates on the ever-evolving wedding plans for Kerry and my big day, in anticipation of one of my mum's increasingly frantic interrogations.

'There's less and less bees every year, they're saying ...'

I walked into my parents' living room just as my dad's latest discourse was gathering momentum. 'They're saying' has long been my dad's method of revealing his sources, an attempt to ensure his heavily paraphrased, embellished monologues hold some integrity when placed under scrutiny.

My mum looked happy to see me; she always does, but she looked especially happy as my entrance offered a natural out, a temporary escape from my dad and his concerns about the dwindling bee community.

A routine visit passed, but a few days later, a hysterical phone call from my mum revealed the true extent to which my dad had enlisted himself as a guerrilla warrior in the fight to save the bee. He'd been out on the top step having a cigarette, one of the twenty highlights of his day, when he saw a bee struggling under a sheet of tarpaulin which was placed over a garden chair. In many Scottish gardens during summer time, tarpaulin offers an alternative to carrying the garden furniture to the shed every evening, in anticipation of a torrential downpour. The bee was released safely, but my dad, aged seventy-two, had over-stretched himself, lost his footing and fell down the original five stairs before rolling and then beginning a rapid descent down the 'big stairs', which lead out onto the pavement. A fall which would have been hilarious to onlookers if he were still in the optimum age bracket of comical falls, approximately between the ages of twelve and the early fifties. It's one of the first signs of ageing, when you have a slapstick tumble that elicits panic and genuine concern as opposed to helpless laughter.

My mum was preoccupied with preparing a ham salad and oven chips – staple Scottish summer garden food – and had missed the incident, only appearing from the doorway to see my dad lying face down and unable to move or make a sound. Being a neurotic worrier and a lifelong doomsday prepper, and also a devout Catholic, her initial reaction was to shout to her best friend and neighbour, Bertha, who was running down her own stairs having witnessed the whole thing – from the bee salvage right through to the fall – to phone their parish priest. First aid, paramedics, nurses, doctors were all cut out of the rescue bid as my mum had pronounced my dad dead and it was now a matter of having his soul redeemed and his last rites read.

The feeling of frantically racing down to my parents' home after my mum had hung up the phone is still vivid and the morbid feeling of helplessness will never leave me, but it was also assuaged (another one of my new words there, folks!) by a rational voice in my head reminding me that my mum is prone to a hyperbolic overreaction. When I arrived at their garden, my dad was conscious and repeating his oft-said mantra: 'Patricia, fucking calm down.' These will no doubt be my dad's last words, but not this day. He was clearly in agony, a little embarrassed and unable to move, but he was already beginning to see the funny side and proudly announcing the news of the bee's emancipation.

My dad has suffered from rheumatoid arthritis for as long as I can remember and had recently been struggling with pneumonia, so it was clear, despite his protests, that he needed medical assistance. It was a serious situation, but maybe not serious enough to justify the amount of missed calls from my mum on the Lord's phone. After a bit of convincing from my mum, Bertha, myself, Kerry and my brother John, and following advice from Scotland's ever-excellent NHS 24, it was decided that, based on the likelihood of serious injury and his recent medical past, my dad was to get over to the Queen Elizabeth Hospital in Glasgow. Grudgingly, fearing being kept in overnight or for a few days, my dad was helped into my car and off we went.

We were seen fairly quickly, given my dad's age and the colour he had turned as the adrenalin wore off and the pain kicked in. The nurse, Sue, introduced herself and calmly reassured my mum that my dad was in the best place and all the standard stuff. She had recognised me, which is always handy in these situations, and it turned out that her brother was the one and only Tommy Flanagan, star of one of my favourite films, *Ratcatcher*, and a major Hollywood player, starring in *Sons of Anarchy* and a shitload of other acclaimed shows and movies. My dad, who was now perking up again, realising a fresh set of ears was now present to hear of his bee rescue, began to rattle off his Netflix recommendations and apologised for having not been able to 'get into' *Sons of Anarchy*, but that he would as soon as he was out, perhaps seeing this as a bargaining tool. An 'if you let me leave, I'll watch your brother's show' type of arrangement.

With the nurse and doctor fearing that the impact from the fall could have affected my dad's lungs, which wouldn't be ideal for his steady recovery from pneumonia, he was taken for an x-ray which, fortunately, came back showing only a few cracked ribs.

My dad was advised to stay in hospital for the evening, but managed to negotiate an early release with a dose of heavy painkillers and a promise to rest in bed and watch *Sons of Anarchy*. He managed a dance at our wedding, even mimicking a few shotguns being fired into the air as the band played their George Ezra cover.

On behalf of my family, thank you, to Sue (Tommy Flanagan is not the only superstar in his family) and all of the staff, nurses and doctors at the Queen Elizabeth who looked after my dad in this instance and countless others in so many other, more serious instances.

We're as indebted to you all as that bee should be to my dad.

In these times, I don't have to detail the worry we all feel for our parents, grandparents and any loved one who fits the high-risk category. So, another thank you, to the NHS staff, across the board, for their expertise, their professionalism, their compassion and their unyielding work ethic in what is a truly overwhelming time. They are the frontline heroes in this war effort.

Thank you also to the staff, nurses and doctors who care for the young people at the Queen Elizabeth teenage cancer wing which I proudly opened a few years back. To everyone at Clydebank Health Centre who has been there for me since I was born, from my MMR jabs to my sore arse, and a special thanks for the care that they continually provide for my parents.

Finally, to Dr Sadozye and Dr Siddiqui from the Beatson Cancer Centre in Glasgow for their dedicated treatment of my mother-in-law, Joanne. I hope this little contribution can in some way reflect my gratitude to the NHS and my hope that they all keep strong and are rewarded for these monumental efforts.

I'll be pouring one out for you all on Thursday night.

MARIAN KEYES

On Monday morning, 17 January 1994, I awoke in my flat in Maida Vale in London.

I'd had a terrible weekend, which had been spent drinking and contemplating suicide. Now it was Monday morning and I was deep in the horrors. My friends and colleagues had been telling me for some time that I was an alcoholic and that I needed help. But they were wrong. I was simply depressed and alcohol wasn't my problem, instead it was the solution.

However, on this grey January morning, I understood intuitively that alcohol was making me miserable, that it had stopped all forward propulsion in my life and it ruined anything good and decent.

I needed to stop. But I knew couldn't. Suspended between those two impossible extremes – keep drinking/stop drinking – it was clear that I couldn't keep on living. There were sleeping tablets in my bedroom, and some antidepressants. I took them all and waited to die.

As I drifted in and out of consciousness, I wondered if there was any other way of approaching this? So I rang a friend, who rang an ambulance, which arrived a short time later.

I remember the paramedics, two men, friendly and brisk. Cheery, even. Strapping a woman into a stretcher before 8am on Monday morning seemed to be nothing out of the ordinary for them.

They took me to St Mary's Hospital in Paddington where I was seen right away. Instead of having to have my stomach pumped, I was fed activated charcoal, which – I subsequently discovered – is how most overdoses are treated.

There are many people who think that taking an overdose is an act of self-indulgence, but none of the staff at St Mary's seemed to think that way: they were so incredibly kind to me.

The act that I remember with the most gratitude is when my poor mother rang from Ireland. These were the days before mobile phones so the nurses let me use the phone on their desk to console my devastated mother. Somebody got me a chair to sit on and everyone continued working all around me as I let myself be persuaded to go to rehab.

The following day, I flew back to Ireland and two days later checked into a treatment centre. I haven't had a drink since.

Because of the way I'd been living, if there had been a cost for calling an ambulance, I wouldn't have done it,

I couldn't have afforded it. I would have let myself die. Likewise, with the hospital care.

Three months later, I returned to London and was able to restart my life, living in a better, healthier way.

I'm so profoundly grateful to the NHS for saving my life and doing it with such compassion and kindness.

NAOMIE HARRIS

I was fifteen and filming *Runaway Bay*, a children's TV series set on the Caribbean island of Martinique. I was filming with five other kids all around my age and this was the third year in which we had the privilege of spending our summer break simultaneously working and holidaying on the glamorous island. However, this year, unlike the others, I wasn't having fun. My mind was entirely consumed with the 'event' I knew was waiting for me as soon as I got back to the UK.

I had worn a back brace since the age of twelve to try to correct my scoliosis (curvature of the spine), but recently the doctor giving me my yearly review had told my mother and me that my scoliosis had continued to advance and I therefore needed to have the Harrington rod operation if I didn't want my breathing to be impacted as an adult. The Harrington rod operation would involve the removal of a rib on the right side of my body, deflation of my right lung to access my spine and then the insertion of metal clamps along my spine to try to prevent the curvature from getting any worse. I would have to be in hospital for a month. Words like 'devastated' and 'terrified' don't even come close to expressing how I felt at fifteen, never having spent a night inside a hospital before, let alone had an operation.

I arrived at the Royal National Orthopaedic hospital in Stanmore, London, sun-kissed, but totally unrested from my time in Martinique. I had a suitcase packed for my month-long stay and a bundle of nerves held in the pit of my stomach. Needless to say, I did NOT want to be there. As my mum settled me into the ward in preparation for my first night alone in hospital, I made a silent pact with myself that I was going to be out of that hospital in *way* under a month. I was planning on being the fastest-healing Harrington rod patient in history!

I have so many memories from my stay in hospital (which did in fact end up being a month) . . . The little boy in the bed across from mine who was having the same operation but had to spend double the amount of time in hospital because his parents were Jehovah's Witnesses and so wouldn't allow blood transfusions; the frustration of not being able to do basic things, like walk or sit up on a chair; the huge Edwardian windows that flooded the ward with light; the smell of disinfectant; the challenge of trying to sleep in a ward full of children making noises all night; my mum having to work but always managing to be by my side. But the strongest and most impactful of all of these memories are of the compassionate faces, encouraging words and embraces of the incredible NHS nurses.

I'd arrived at the hospital full of bravado, desperately trying to hide the terror I felt about my impending operation from my mum and everyone else I knew. I'd managed to fool a lot of people – they called me 'brave' and said I was 'coping so well' – but my act didn't fool the nurses at Stanmore for a second. On my first night, as soon as my mum had said goodbye and headed home, one of the nurses came and sat on my bed. I can't remember exactly what she said, but it was something to the effect of, 'You know, Naomie, it's OK to be scared, every single child here was afraid before their operation.' It wasn't really so much what she said that affected me, but the insight with which she said it. It was like she understood what I was feeling without me having to say anything, and she gave me full permission to let down my guard, to stop pretending to be stronger than I actually was and to allow myself to be a scared kid. That nurse will probably never know just how powerfully healing that moment was for me.

There were so many other moments like that with different nurses that I vividly recall from my time in hospital . . . The love wordlessly communicated to me after my operation through the care with which nurses cleaned and tended to my scars; the patience with which my arm was held as I was guided around the ward as I struggled to learn to walk again; the gentleness with which I was helped to wash and clothe myself while I healed, and the many, many moments when I cried and was held and comforted.

Nearly three decades later, I could look back on that period of my life as grim and challenging, but because of the compassion and care I was shown by those NHS nurses I remember it instead with gratitude. At a time when I had no choice but to let my guard down and put my faith in others, I was shown that there are so many people in this world who are kind, who care and who are there to help. It's one of the most powerful lessons I've ever learned. To me, those nurses were as close to living, breathing angels as I've got and I will be forever grateful to every single one of them.

CHRIS O'DOWD

When the Heart of a Man Meets the Leg of an Elephant

I had been feeling well, as it happened. I was twenty-seven, carrying a few unnecessary pounds, a couple of which I had just picked up through an overly indulgent visit to a local kebab establishment. Otherwise, I was in fine fettle, I presumed, like a fool. I ambled along merrily from Stockwell tube with nothing but a song in my head, a spring in my step and a whistle in my pocket. (I always carry a whistle in my pocket in case any ad hoc refereeing needs doing.) But my merry meander was cut short, as, from nowhere, my chest tightened, my legs bore a knuckle and my stride became a stumble – my robust frame collapsing slowly like a poorly constructed flan.

I was lying on the footpath now, which wasn't ideal. I was sweating, though it was a December evening in London, which also seemed unhelpful, at the time. As my human form slowly slumped across the concrete paviours, my left arm tingled in morbid excitement. The flapping flutter of my limb was understandable. It had never witnessed a heart attack before, let alone shared a body with one. Yes, readers, my heart, long my burden, was giving way at the tender age of twenty-seven. I had attracted the 'rockstar curse', it seemed, which was fittingly above my station. Even my heart lacked the humility to retire at a less notorious age. I was alone, phoneless and at the mercy of the commuting Samaritans of south London.

A passing skateboarder (Stockwell has a thriving skateboarding scene which I hope to embrace and indeed rule one day) came to my aid. I breathlessly asked him to call me an ambulance. He nobly sidestepped the comedic landmine and actually phoned for help.

The London Ambulance Service came swiftly, as is their wont. The dreadlocked paramedic asked me how I was feeling. He took my pulse and made me comfortable. He offered to contact someone to accompany me to hospital. (I was soon joined in the mobile health wagon by my ex-girlfriend, or 'the one that got away', as I've been told to stop calling her at home.) It's worth noting that at no point during any of this did the paramedic make any mention of payment or insurance plans. Nor did he question my nationality or my means. In contrast, we live in California right now, where once my wife, deep in labour, was stopped from entering our delivery ward until all the correct paperwork was signed and filed. She stood at the reception feeling the baby crown like an evil prince until we were seen as financially fit enough for a bed.

Not in Stockwell though. That night, they gathered me up in their caring embrace, like an older sibling. 'Don't worry, pickle,' they seemed to say. 'We all have bad days, we'll help you fight another!' the sirens screamed.

It was mercifully quiet at Guy's hospital in London Bridge. It was a Tuesday, I believe. My nurse's name was Tina. She was Scottish and it's true what they say, they're just like normal people, the Scots. Tina was supremely soothing but took zero shit from yours truly. She recognised me from the telly but insisted the other fella on the show was the funny one, a point to which I reluctantly agreed. In some feverish attempt to bond with my carer, I informed Tina that my mother had trained as a nurse here in the 1960s and asked if maybe she knew her. Tina coolly informed me that she was fifty-six. So I changed the subject.

Doctors came and went. The A&E staff, who seemingly use their free time to take plate-spinning classes, lurched in and out of my 'situation'. They strapped some wires onto my muscular torso and six-pack. There was an ECG machine, I seem to recall, but I'm no letter-expert. Assessments happened and I waited, all the while writing an unimpressive will in my head. As well as the first draft of my own eulogy.

Before too long, Tina returned. With a warm smile, and a hint of a smirk, she said, 'We'll keep an eye on ya Chris, but we're fairly sure it's just trapped wind.'

The air went out of the room. Probably looking for its friends, who were congregating within my chest cavity.

'It's not my heart?' I asked, almost disappointed for some reason.

'Your heart is fine, just go easy on the meats, there's only so much room in there,' she replied, her tone warm, her eyes already moving on to a patient in more need.

As I left the hospital, I wasn't handed a bill that would cripple me but rather a pamphlet on healthy eating, which, of course, I ate immediately.

If there's a lesson here, it's probably the importance of exercise and that a meal served from something lovingly called an 'elephant's leg' should probably be enjoyed in moderation. If there's a moral, it is the glorious consistency of healthcare professionals. Even for idiots like me, they are there when we need them, and they are there when we don't.

Thank you NHS, now, and always.

ED SHEERAN

I, like most people in the UK, grew up with the NHS. I was born in Halifax in an NHS hospital, was in and out of the GP surgery every few weeks with ear infections from my first few months, had laser treatment for a port wine stain on my eye at Leeds General Infirmary over several years, had stitches on various parts of my body from always being a curious kid all throughout my childhood, had my eardrum replaced when I was eleven in Ipswich Hospital's children's ward and got both arms put in casts in Ipswich A&E when I fell off my bicycle mid-tour in 2017. And I haven't even touched on dental and eye care! The NHS really has been there through every single up and down of my life so far, and every visit – even though I was either in pain or terrified – wasn't ever as bad as I was thinking it was going to be, and that was down to how professional and lovely the doctors, nurses and support staff were.

Although I am grateful for all of the times the NHS has been there for me personally, the thing I'm most grateful to them for is the care they gave my grandmother in the final few months of her life. She was in Aldeburgh Hospital in Suffolk with terminal cancer but had had a lifetime of health problems that left her in constant pain. My grandfather, her husband, worked as a doctor in the NHS for much of his life, but that's another story.

I was lucky enough not to be on tour during my grandmother's final months, and because I lived locally I was able to visit her every few days. The care she received was incredible; the people who worked there so lovely, compassionate, funny and caring. When she passed away, I wrote a song called 'Supermarket Flowers' about the situation. The verse lyrics are about packing up her room at that hospital. Me and my family became very close to the nurses who worked there and my mum is still in touch with them now. I see them from time to time when I'm in the area and it's like meeting old friends.

Places like Aldeburgh Hospital just don't exist in large parts of the world and, in many, places like this are private and cost a lot of money, or you have to have health insurance to be able to access them. The NHS is unique. It can be taken for granted, or just accepted as the norm, but it's not the norm. Without sounding cheesy, it's the backbone of our country and idolised by me and millions more. This time of crisis has reminded so many how important it is to us, and to the country, and I hope, going forward, it will get the financial support and respect it deserves. It was created for a reason – to provide consistently excellent, reliable healthcare to all, regardless of background or means – and we mustn't ever lose it.

Thank you to everyone who works for the NHS. Thank you for putting other people's pain and suffering above your own, and thanks for all you've done for me and my family.

DAME JACQUELINE WILSON

When my daughter Emma was small there was a special book she wanted me to read aloud every single night. It wasn't *The Tiger Who Came To Tea* or *Where the Wild Things Are*, though these were great favourites. It was a little Ladybird book called *The Nurse.* It was a simply written, rather prosaic non-fiction book about nurses, with a picture on every page. Emma soon knew the text by heart and could chant along with me.

She thought the 1960s nurse's uniform very glamorous. Certainly many people have a nostalgic yearning for the prim blue dress with its tight elasticated belt, starched white apron, elaborate cap and shiny black stockings, though not always for the right reasons. Emma decided she desperately wanted to be a nurse herself and played endless games with her toy nurse's kit, taking it all very seriously.

Then one day a small friend came to tea. This child ate eagerly enough – but almost immediately was violently sick, all over herself, the table, the carpet, even the wallpaper. Emma retreated to the other end of the room, horrified, while I did my best to comfort our guest and start the much-needed mopping. I'd done the same for Emma before, never a pleasant task, but somehow it's worse when it's not your own child.

Emma watched, pulling a disgusted face.

'Wait till you're a nurse. I expect you'll have to clear up sick lots of times,' I said, meanly.

She was shocked. 'Nurses don't have to do *that!*' she said.

'Oh yes they do,' I insisted.

Emma nodded and hid the Ladybird nurse book in the cupboard.

I've just been looking through a copy of it, shaking my head wryly at the dated explanations, but in some ways the words still ring true: 'Nurses in hospital work very hard. Although the nurses may sometimes feel tired, they are always cheerful and smiling.'

Smiling is probably the very last thing anyone feels like doing in the NHS at the moment, and it's hard to see lip movement behind a mask – but they're certainly working harder than they ever have in their lives.

I learned to appreciate the NHS long before we knew there was such a thing as coronavirus. For the first sixty or so years of my life I was in perfect health and I'd only had brief spells in hospital for mundane reasons: tonsils and adenoids as a child, childbirth as a young woman and a small lump removal when I was around forty. I took pride in my fitness, my energy, my stamina, producing two books a year and packing in countless visits and events and long signing sessions.

Then suddenly my body rebelled. First it was heart failure, sudden and dramatic and life-threatening. I shall forever be grateful to the doctor who first diagnosed me and the medical team at the Royal Brompton Hospital. It took a little while for me to recover and to get used to my own private defibrillator sewn inside my chest, but I felt I'd got my Big Medical Drama over and done with.

I didn't realise that my kidneys were now failing too. Within a few years I had to go on dialysis, not the most pleasant of processes, but much better than the alternative. I grew to love the gentle, efficient nurses I saw three times a week, mostly from the Philippines, the kindest and most cheerful medical staff ever.

I was lucky enough to have a transplant after a couple of years and the renal team at St George's were wonderful, both to me and my donating partner. There was one funny moment when the young nurse admitting me needed a swab from my groin to make sure I wasn't harbouring any bugs. I was standing with my jeans round my ankles, trying to look nonchalant, when she looked up and squinted at my face.

'Oh my God, you're the Tracy Beaker author!' she said. 'I love your books!'

It's always lovely to meet a fan, but perhaps it's better when you're fully dressed!

All the staff on that unit were incredible. I have particularly fond memories of the soft-voiced star of the tea trolley, a young woman who served the food on our ward. The morning after my transplant, she came tip-toeing into my room and woke me very gently, with a cup of tea and a slice of toast spread with butter and marmalade, cut into squares as if I were a little girl. I don't think anything has ever beaten that breakfast.

You don't sign off completely after a transplant, so I report back at regular intervals. I have been known to fret about the hours waiting to give blood, or the endless weighing and monitoring. Never again! I'm newly grateful to the NHS, for the time and care and money they've spent on me. They've literally saved my life twice over and they're currently risking their own lives to nurse every patient that they can back to full health.

God bless the NHS!

PAUL McCARTNEY

As far back as I can remember, my mother was a nurse. She became a sister on an NHS ward and later was a midwife, so my feelings about the NHS are obvious – it's a fantastic institution. I know that anyone who works for them sees it as their vocation and is a real hero.

One of my outstanding memories is that, one night, in the winter in Liverpool where we lived in the midwife's house, my mother was called out to assist a local lady in giving birth to her baby. I remember standing by the front door watching her leave the house on her midwife's bicycle with a basket on the front and her medical case on the back. The road was covered in a good few inches of snow, but she had no alternative other than to go to the birth in her midwife's uniform on her standard-issue NHS bicycle. She set off leaving tyre tracks in the snow.

That moment will be with me forever and encapsulates my pride in, and gratitude for, the National Health Service. We are so lucky in the UK to have such dedicated people to look after us all and no matter who you are you can still benefit from this fantastic system. Thanks NHS. Thanks heroes. Thanks Mum.

Love, Paul

MONICA ALI

Mrs Antonova

A deep stillness had settled over the ward. The high winds of washing, dressing, toileting, consultant rounds, medicine rounds, food trolleys, physiotherapy, occupational therapy, bingo, Scrabble and hairdressing had subsided, leaving the place eerily becalmed. Yasmin looked around. The linen cage was almost bare, only a few white sheets and pillowcases hanging limply over the wire racks. The door of the equipment room stood open to reveal a leaning tower of commodes, loosely piled hoists and a jumble of drip stands. Yellow holdalls of dirty linens lay like sandbags in front of the pharmacy cupboard. Nothing moved.

Visiting hours were often like this. Yasmin tiptoed past the bodies, the closed eyes and open mouths, the rows of arms pinning the bedclothes as if rigor mortis had set in already. She hovered at a bedside, fighting the urge to poke and prod. There was no need to go round checking everyone was still breathing, disturbing everyone's rest.

It was only exhaustion. Old age and sickness and sheer exhaustion, because washing and dressing started during the night shift at 5.30 a.m. If there were no visitors during visiting hours it was perhaps a blessing because the patients could get some sleep.

Something stirred. Yasmin turned to see Mrs Antonova struggling to sit up in bed.

'Let me help you.' She rushed across. Mrs Antonova's ribcage felt so fragile beneath the brushed cotton nightdress, when Yasmin put her hands on her back.

'Thank you, pumpkin.' Mrs Antonova called everyone pumpkin, from the cleaners to the consultants. She was ninety-six years old and this was her third stay in hospital in as many months, this time following a fall at home. 'Now, if you'd care to do me another favour, just get one of those pillows and hold it over my face. Don't let go until you're absolutely sure.'

'I'm sorry you're feeling that way,' said Yasmin. 'I can prescribe you an antidepressant to help if you're feeling so low.'

'I'm sure you can, pumpkin. But I'm not depressed.'

Yasmin sat down on the bed and took Mrs Antonova's wrist. Her pulse was a little thready but nothing out of the ordinary for a patient with chronic atrial fibrillation and arrhythmia.

'I'm not depressed, I'm bored,' said Mrs Antonova. 'I'm surrounded by ...' She peered around and Yasmin was struck by her regal bearing, even in her rucked-up nightgown, with her wig askew. 'By *that*!'

'Mrs Antonova,' said Yasmin, 'do you mind if I sit here with you for a while?'

'Call me Zlata.' Mrs Antonova, since her readmission, seemed devastatingly frail. Impossible now to imagine her being strong enough to push a trolley. Which was how she'd wedged the door handle of the television room for the protest she'd staged during a previous stay. But her voice was still full of mischief, and she winked at Yasmin as she said, 'Man trouble, is it?'

'Oh, no,' said Yasmin, smiling, 'nothing like that. Just thought we could keep each other company.'

Mrs Antonova tugged at her wig, making it even more lopsided. The aubergine-coloured curls, thick and shiny, contrasted impressively with her paper-thin, paper-dry skin. 'Man trouble! I know it when I see it. Married five times, and so it goes.'

'Five times!' said Yasmin. Mrs Antonova, on a previous stay, had told her about three husbands: a Uruguayan dentist who turned out to be homosexual; an Israeli violinist who had spent two years in Treblinka and who, three days after a honeymoon at the Sea of Galilee, committed suicide by jumping in front of a train at Tel Aviv Savidor Central (it was the 9.45 p.m. to Hod HaSharon) and a civil servant from Bexley Heath who, Mrs Antonova indicated by a vague wave of the hand, was not worth talking about.

'Yes, five. Do you know how old I was when I was first married? Sixteen. That's eighty years ago.'

'Goodness! I'd love to hear all about it one day.' She should be catching up on paperwork while the ward was quiet. She should be snatching an opportunity to study for the MRCP exam or update her reflective practice on her e-portfolio. She was behind with everything.

Mrs Antonova leaned towards Yasmin. It was a risky manoeuvre and for a moment Yasmin feared she would topple sideways out of bed. 'You are very busy, pumpkin. Thank you for the visit. It was lovely.' She sounded like she meant it. Mrs Antonova looked her age, or rather, she had moved beyond the age where people guess an age ('Doesn't he look great for eighty-two?') into a fearful zone of decrepitude that tends to evoke pity, not admiration. But her voice had not aged with her; it was strong and singsong. Playful.

'I'm not busy,' said Yasmin. 'Tell me about your first husband.'

'Dimitri Ivanovich Shestov was fifty-three years old when I married him, and I had just turned sixteen. Of course it wasn't my idea to marry Dimitri, I had no say in the matter. I believe my father owed him money or something silly like that.'

'How awful.' Yasmin felt sorry Mrs Antonova's only 'visitor' was a doctor. She felt sorry for the entire somnolent ward. The cancer wards, at visiting time, turned into a scrum. Cancer made you popular.

'He was the love of my life,' sang Mrs Antonova. Somewhere in the rubble and ruin of her body, Zlata, the sixteen-year-old bride, still lived. 'He was a white Russian, an émigré, like me – do you know about the white émigrés, pumpkin?'

Mrs Antonova explained that she was a baby when her parents fled Moscow in 1921. Her father, Vladimir Antonov, was an academic. Her mother, Nataliya, was only twenty years old when they fled to Istanbul, and eventually Prague, Paris, London. She took with her baby Zlata, a Kornilov tea set and a determination to live in exile the life of a Russian noblewoman. The Kornilov teacups had little gold griffins for handles. The insides were fully gilded and the outsides were painted with seashells and flowers. 'You know what we were, pumpkin? We were gypsies. Noble gypsies roaming Europe with our teacups and rye bread and Russian pride. Sometimes we had no bread. But Vladimir had his writing and Tashenka had her cups and they both had their pride. Pride, you know, is a very expensive commodity.'

Mrs Antonova fell silent.

'So what happened with Dimitri?'

'Who?' said Mrs Antonova, closing her eyes.

'The love of your life,' whispered Yasmin.

JIMMY CARR

Dear NHS,

Not all stories have happy endings. Spoiler alert: in this one, my mum dies. In spite of advances in medical science, miracle cures, Hail Mary passes and last-minute reprieves, the reality is, for all of us, death is our ultimate fate. But thanks to the NHS, none of us have to face it alone.

My mother Nora Mary Carr (née Lawlor) was a nurse. She trained at the Regional Hospital in Limerick, Ireland, and came over to London in the early 1970s. She was one of those migrant workers, the ones who (when we're not having a global pandemic) get given a hard time: 'These immigrant doctors and nurses, they come over here, saving our lives.'

My mother was – how can I put this? – fucking hilarious. Any talent I have in me, I got from her. I don't know what she'd make of my stage act but I think she'd want a credit and royalties. Fun, loud and inappropriately sweary, she was the life and soul. She had a literally breathtaking laugh.

If you really got her good, she'd fall completely silent, eyes half closed as she slowly rocked back and forth, looking like she was having some sort of fit. To be clear, this isn't what killed her, but what a way to go.

Some say: 'Grief is the price we pay for love.' And some say: 'Home isn't a place, it's a person.' Well, my mother was my home and, when she went, the grief pretty much broke me. She was young – in her mid-fifties – when she died in Guy's and St Thomas' Hospital, London. Opposite the Houses of Parliament on the River Thames, you could scarcely find a more picturesque spot to watch a loved one fade away and die.

Twenty years later, and I still feel the waves of grief. I'll find myself driving the route I took to visit her in hospital and it'll hit me. I'm right back there in 2001. She passed in September, just before 9/11. It felt like the sky was falling and the world was ending – just like it does now.

My mother first got sick in January of that year. That's nine months of 24-hour care, day in, day out. There are countries in the world where that simply could not happen. Who would have paid for it?

The care my mother received was, as you'd imagine, exemplary. But it was not just my mother who was looked after in that hospital. Through their support and unending kindness, the doctors and nurses also took care of my brothers and me, and helped us to cope. Or to pretend to cope, which I think is as close as you can get in that sort of situation.

Pancreatitis isn't a fun way to die. There are false dawns; you're in and out of intensive care. It must have been hard for my mother to get the diagnosis – she'd have known the prognosis all too well. I remember at the end, a nurse told me that I should call my brothers. She said that my mother had around five hours. Imagine that. The nurse had seen so many people die that she could say, with accuracy, what was coming and when. Mercifully, she knew the signs, which allowed us all to be there at the end.

If you can be with a loved one when they die, you should. Her hands getting cold as the circulation shut down, her breathing getting heavy, the death rattle. Bearing witness to a death is an incredibly intimate thing. You should be there, not because it's easy – it isn't – but because one day you'll want someone to hold your hand.

It's amazing to me that I can look back on a period of my life that ended with the most important person in my world being taken away with a prevailing feeling of gratitude. Gratitude for the extra time we got with her. And gratitude for the love our family were able to share with her in an environment that felt constantly supportive. The truth is, in those nine months that hospital became our home and the NHS became part of our family.

Of course, you want to go back to thank the doctors and nurses who fought the good fight. Who allowed my mother to die with dignity and, thanks to the pain management team, peacefully. But I can't go back because that ICU is gone. It's still there, obviously, but those doctors and nurses are no longer on shift and someone else's relatives pace the halls. And it's not like I have one doctor and a couple of nurses to thank. It's an entire system. I've got to thank ALL of them. The doctors, the nurses, the hospital porters and custodians, the administration staff – there's an army of people who make up our NHS.

'National' is at the heart of the NHS, itself at the heart of being British. I've said it before and I'll say it again, perhaps we should take the politics out of the NHS. It's just too important to be a political football. Maybe we should do what they did with pensions. Triple lock it, link the spending on healthcare to GDP and write it into law. Be generous. Let's take care of the NHS before we think we're going to need them.

And make no mistake: we will all need the NHS. Private healthcare is all very well for ingrown toenails and Botox – it's for vanity. But anything serious, you want the big guns. You want the NHS. If you've got cancer or Covid-19, you don't give a fuck about a colour TV in your room.

Not all heroes wear capes. A lot of our NHS staff don't even get to wear the proper PPE. It would be remiss to not mention and honour the healthcare workers who died caring for us in this pandemic. The unimaginable bravery they knew. Like firemen walking into a burning building, they knew the risks and they worked anyway. So many of them did not make it out.

I have no god. No disrespect if you do, 'May your God go with you,' as Dave Allen would say. Prayer to me is abstract; it's whispering into the abyss. But I do believe in the NHS. I know if I dial 999, an ambulance is coming. It's real. I know if I walk into an emergency room, a doctor will see me. They're always there.

It's hard to think about the future in the middle of a pandemic. But one thing I do know is that, one day, odds are I will die in an NHS hospital. And I'd like to say thank you. Not only for taking care of my mother but, as I'll not be in a fit state to say it on the day, thank you for being there to comfort and care for me at the hour of my death.

BENJAMIN ZEPHANIAH

Praise the saviour

In Barbados, my grandfather, a twin
But his twin brother died at birth.

My father, he too was a twin
But his twin brother died at birth.

I am a twin,
A difficult twin.
And I almost died at birth.

I was an emergency.
Born in a corridor
Of the hospital that my mother worked in.

I never saw my father cry,
But I'm told that he cried when he saw me
Struggling to enter the world via Birmingham.
I'm told that he cried when he saw my mother
Struggling to give birth to this dub poet

The first of a set of twins,
Already growing dreadlocks.

But this was England, and
I made it.

Twelve I was when they found a lump in my breast.
'But boys don't get breast cancer.' My father said to the doctor.
I thought I was going to die, but
I made it.

Fifteen I was when my lungs collapsed.
An athlete with no lungs won't work,
And I thought I was going to die, but
I made it.

Twenty-five I was when my testicles blew up.
I told my girlfriend it was a football injury.
Told macho lies to the guys. But
I made it.

Fifty-five I was when in the middle of a lecture,
In front of my students
In between Homer and Ginsberg,
My appendix burst. Pop.
Yes, I thought I was going to die. But
I made it.

I went down the tunnel, I saw that light.
And they brought me back. So
I made it.

I made it because someone had a dream of healthcare for
all.
I made it because someone said this is for me, and free.
I made it because some idealistic rebel thought I deserved
it.
So, when I think of why I'm here, I think of it,
When I think of all my mother's work, I think of it,
And when I go into that little room, I vote for it.
I've loved it since the day I was born.
The NHS is for life, not just for emergencies.

JOHN NIVEN

Rock Biographies Lied to Me

It is the early nineties. I have just graduated from university and am living with my then-girlfriend Stephanie in a one-bedroom flat in the west end of Glasgow. We are returning late one night when I see that our neighbour Calum's door is wide open. Calum is in his fifties and very much a loveable rogue. He's a Glasgow journalist of the old school: a hard-drinking bon viveur who is never happier than when he's typing up a no doubt heavily padded expenses claim on the battered Amstrad computer that sits on his desk in his front window. Calum is prone to the usual stuff that accompanies his lifestyle: locking himself out of the building and so forth. 'You go on up,' I say to Steph. 'I'll just check he's OK.'

I knock lightly on the open door and, getting no response, wander into the hall. 'Calum? Calum?' I shout. 'Your front door's open.' There is noise and movement coming from the bedroom. I walk in.

On a chair in the corner sits a fairly rough-looking Glasgow ned in his thirties: crew cut, shell suit and dirty trainers all present and correct. He looks somewhat … glazed. Calum is laid out flat on his back on the bed. Standing over him, frantically slapping his face, is his elder brother, Duncan.

Duncan is an equally hard-liver – a fisherman from the north of Scotland (where both brothers come from) who occasionally visits Calum in order to tear up the town. 'Ah, hey Duncan,' I say. He stops slapping his brother and turns around. 'Everything OK?'

Everything is not OK.

Some backstory here that I already knew about. Calum had recently been researching a story about drug gangs in Glasgow and had got to know some lads like our shell-suited friend in the corner. Over the last few months, he'd gone from just being a boozer to occasionally doing some cocaine as well to, more recently, asking me if I can get him some Es now and then. His Jimmy Corkhill-esque trajectory has reached its apotheosis in the scene now being played out in front of me. Duncan fills me in.

Earlier that night, they had been drinking with some rough lads Calum had gotten to know when he decided to front Shell Suit the money to get some heroin in. They'd retired here to the flat where Shell Suit shot up Calum and then himself. Duncan – sensible old fisherman that he is – stuck with the whisky. This was not long ago and apparently Calum is 'no handling it well'.

I go over and my first thought is that this is correct – he's dead.

Calum is chalk-white, his eyes closed. I prise open an eyelid and am astonished to see something I have only read about in rock biographies – the phenomenon of being 'pinned'. His pupils are *microscopic*, tiny little pinpricks. His chest is barely rising and falling. He is breathing. Just. 'I . . . I think we need to call an ambulance,' I say.

Finally, Shell Suit speaks up. 'Naw man. Fuck sake. You'll get us all the jail. He'll be awright.'

Of course, with hindsight, I should have called the ambulance there and then. However, I don't want to get Calum into trouble. I don't know how it works when an ambulance crew arrives at the scene of a heroin overdose. Do they automatically call the cops? I'm in my early twenties and massively out of my depth here. I do know one thing though: he is very, very far from being 'awright'. 'We need to wake him up,' I say.

'I've been trying that for half an hour!' his brother replies. Indeed, I see now that Calum's hair and shoulders are soaking wet from the water that has been thrown over him. His cheeks are reddened from all the slapping.

It occurs to me that Duncan is paralytic and Shell Suit is smacked out of his mind. Early twenties or not, I'm the only adult in the room. It is now that something else comes back to me from the yards of rock biographies I have consumed, something that I read about in connection with Gram Parsons, the country-rock singer who died of a heroin overdose in a Joshua Tree motel in 1973. 'Right, strip him off and get him in the shower,' I say. I run into the kitchen and root in the freezer until I find the ice-cube tray. I run through to the bathroom where they have the naked, still unconscious Calum laid out on the floor of the shower stall. We start running it as hard and as cold as we can get it. Nothing. No response. It has to be done.

I begin trying to insert ice cubes into Calum's rectum.

Have you ever done this? Either out of necessity or for pleasure? (There are more things in heaven and earth and all that, Horatio.) It is not as easy as you might imagine. After a few cubes have been refused entry to his tightly puckered anus and have gone skittering uselessly off across the lino, I realise that my instinctive technique – to try and press the cube home with my thumb – is defective. Better to slightly spread the cheeks with one hand, place a cube in the palm of the other and gradually try to work it home.

Of course, the process is nowhere near as calm and measured as I make it sound here: there is the cold, pouring water soaking me, Shell Suit laughing his head off and Duncan screaming, 'WHAT THE FUCK ARE YOU DOING?!'

Finally, I pop one home and the effect is immediate: for the first time Calum opens his eyes. I ram another one in there. And another. He's awake now. And looking confused. As well you might if you'd drifted off on a rainbow cloud of opium and woken up in a freezing shower stall with your neighbour brutally ramming stuff up your fucking arse.

We get him out and wrapped in a towel and start walking him around the flat, trying to get sips of black coffee into him. Exhausted, wrung out, I say, 'OK. I'm going to bed now. For God's sake, don't let him go back to sleep.'

I head up to the top floor. Stephanie – who already disapproves of Calum as an occasional drinking partner of mine – is asleep. I turn in too, already working on the lightly bowdlerised version of events I'll present over breakfast in the morning. I drift off but am soon woken by Steph elbowing me in the ribs and hissing, 'There's someone at the door!' Not the main door, the buzzer. The flat door. I go and open it to see a guilty, despairing-looking Duncan. 'Sorry John, he's gone again.'

I throw a dressing gown on and head back downstairs. Calum is naked on his back on the bed and, if he didn't look great before, I am now effectively looking at a corpse. I muster all the authority I can and say, 'We're calling a fucking ambulance.'

Shell Suit immediately stands up, nods (my memory wants to add the embellishment that he formally zipped up his shell suit, as though buttoning a topcoat) and bolts out of the flat, down the hallway and off along the street. I ring 999 and tell them we have a heroin overdose. Duncan slumps into an armchair, head in hands, and I wearily retrieve the ice-cube tray for the second time.

I am busy at the coalface once again when I am tapped on the shoulder. I hear the crackle of a radio as I turn to see green uniforms and medical bags. 'Uh, son, what are you doing?' I begin to babble – probably about Gram Parsons in Room Eight of the Joshua Tree Inn – but am simply moved out of the way. The paramedics shine a light into his eyes. They quickly find a vein, produce a preloaded hypodermic of some kind and inject Calum. Two years later, watching *Pulp Fiction* in a dark cinema, I will bark with laughter at the Mia Wallace scene. In reality, it was nowhere near as dramatic. He didn't leap six feet across the room. Calum simply gasped as he sat up, looked around him and said very clearly, 'Hey! What's going on?'

'You've had an overdose. We're taking you to hospital.' Calum nodded. Perfectly content to submit to this plan.

And with that, they helped him off down the hall and towards the waiting ambulance – these men who came out of the night and who saved his life as casually as delivering an American Hot, these heroes who would shortly be on their way to some other hell unfolding while most of us slept peacefully. At the door, one of the paramedics turned around and faced me. 'The ice cubes up the bum thing, son? Urban myth. Just call 999 next time.' It would not be the first or the last time rock biographies lied to me.

And off they went, blue-light circles strobing in the dark.

The cost? Nothing. And god bless the NHS.

The anecdote value? Priceless.

JACK WHITEHALL

I'd never been to a children's ward before, let alone a children's cancer ward, so didn't really know what to expect when I embarked on my first visit to the Oak Centre for Children and Young People at The Royal Marsden Hospital. It was Christmas week 2017 and I was visiting to meet the patients and bring some seasonal cheer.

One of them was a young boy called George, who had just finished his twelve months of treatment (fourteen rounds of chemo, thirty rounds of proton therapy, surgery to remove part of his spine, three muscles and his L3 nerve from his back) for a very rare form of cancer called Ewing's sarcoma. I was fully prepared to be faced with a child at death's door, bed-ridden, semi-conscious and haunted, with doctors talking in hushed whispers and maybe his parents weeping in the corner. So I was shocked to arrive at his ward and find an empty bed. I feared the worst; maybe he'd been rushed in for more emergency surgery, or worse still . . .

'Where's George?' I asked the nurse, bracing myself.

'He's playing football in the corridor with his brother,' she replied very casually. 'I think they need a goalie.'

I went out to find them both charging around the corridor, dodging nurses as they showed off their skills with the football.

'You're late, get in goal please!' yelled George at me, not wanting some grip and grin with a random comedian he'd never heard of to get in the way of his football match.

Well, you can't say no to a child recovering from cancer. His little bald head, the NG feeding tube coming out of his nose and his sallow complexion belied the bundle of energy bouncing in front of me. I took my position between the posts – well, when I say the posts, I mean the door frame that had been repurposed as their goal. A small crowd of parents, nurses and assorted hospital personnel had now gathered.

George stood over the ball and sized up his options. I mentally weighed up mine. I remember thinking, whatever you do, DON'T save this penalty. If you don't let him score, you are going straight to hell.

He'd barely started his run-up when I fully committed to the dive, lying prostrate on the floor, waving my limbs around in the air for a good five seconds, ushering him to roll the ball into the gaping space I'd left for him. Unfortunately, George shanked his kick directly into one of my flailing legs. An awkward moment ensued, as the onlookers tried to comprehend what had just happened.

Did he really just save a five-year-old cancer patient's penalty?

'Encroachment!' I announced authoritatively, pointing at a random doctor who was walking through the ward just at that moment. 'The defender was in the box before the penalty was taken. You'll have to take it again, George.'

His older brother whispered words of encouragement into his ear as he positioned for the retake. The second time I stayed rooted to the spot and George creamed it into the top left corner. The crowd erupted into joyous whoops and applause, relief for everyone. Other than the poor sod who'd have to fix the cracked glass in the door.

One year later, I was back to visit The Royal Marsden and see George again. Thanks to the amazing doctors and nurses he'd made incredible progress, but was still very much in the teeth of his recovery journey. That said, the change in him was amazing, both mentally and physically. A year ago, during the penalty shoot-out, he'd looked like a mini Zinedine Zidane. Now he had a full mop of hair. What hadn't changed was his inability to stay still for a minute. Upon my arrival this time, his mum Vicki – a fundraising colossus who, together with dad Woody and big brother Alex, was in the process of raising £1,000,000 through their George and the Giant Pledge Campaign to help beat childhood cancer – handed me a Nerf gun.

'They're waiting for you in the courtyard. Good luck,' she said, as I headed out to meet my fate. I felt like a young private being sent over the top of the trenches. I walked into the empty courtyard.

'George?' I called. No response. 'Alex?' Silence. They must have got bored of waiting.

I turned around, to be faced with them springing out from behind a bin where they had been lying in wait. It was an ambush! They unleashed a vicious battery of foam bullets, straight at me. I tried to retaliate, only to discover my gun wasn't loaded. I'd been stitched up, I never stood a chance! I got a right royal pasting as their laughter rang out around the battle scene.

These two experiences and several others that I have shared with some of the extraordinary patients whose paths I have crossed on my visits there have helped shape my understanding of The Royal Marsden and probably all NHS children's cancer wards like it. Places which, on paper, should be the most depressing places to visit on earth, but are actually full of not just bravery, courage and tales of extraordinary resilience in the face of terrifying adversity, but also of hope, compassion, love, care, positivity and – most unexpectedly of all – laughter. That's down to both the attitude of the patients and the phenomenal and dedicated NHS staff, who are committed to not only giving those kids the best chance of beating their illnesses, but doing so in an environment where they can still be children. Where they can play, laugh and not let cancer take over their entire lives and define who they are.

Personally, I have been very lucky to swerve any hospital activity, apart from one minor event that meant I wound up in the A&E department of Kingston Hospital, and I have a scar on my knee to prove it. It's one of those scars that over the years I've attached all manner of fictitious stories to.

'How did you get that scar, Jack?' people ask. And my response varies between, 'It was in a street fight,' or 'Skiing down a black run too fast.'

The truth is far less impressive. The reality is that I went to pet a dog, which barked at me, whereupon I ran away shrieking with fear, attempted to jump up onto a wall, tripped and cut open my knee on the sharp edge of the coping stone on the top. When I recount this story I like to elucidate that the dog was a massive bloodhound with an enormous jaw, razor-sharp fangs and a demonic, terrifying bark. My mother, however, will always point out that actually it was a small, cute but yappy, terrier puppy. But I think the gin must have corroded her memory; I'm sticking with my version that this mutt looked like the Hound of Hades. What we can agree on with certainty, is the fact that I was sliced open right down to bone, sinew and tendons and needed prompt hospital attention. And also the date, 1 July 1998.

I remember this date so specifically as it just so happened that my first ever trip to A&E coincided with England playing Argentina in the knock-out stages of World Cup '98. I recall sitting, blood still oozing from my knee, thinking, please don't call my name out, as I sat watching the television that was playing the match, albeit muted, bolted high up on the wall in the corner of the waiting room. I had to be one of the only people in A&E that day, or come to think of it, any other day, desperate for MORE waiting time.

As the game plunged into extra time I was called through to the waiting doctor in the treatment cubicle. After a quick discussion and some hasty rearrangement of the furniture, the doctor manoeuvred me into a position where he was able to do his work, returning the contents of my knee to their rightful positions and stitching me up, and I was able to do mine, gazing over his shoulder and watching the denouement of this epic encounter play out on the waiting-room television. So intent on the game was I, the doctor observed when he had finished, that he needn't have bothered with any local anaesthetic. To be honest, I was so engrossed that I reckon he could have whipped out one of my kidneys and I wouldn't have noticed!

At half-time of extra time, as he put the finishing touches to my dressing, I remember remarking to my mum how unlucky all the poor doctors and nurses were that they had to work during the game. To a ten-year-old boy there could be no greater example of the sacrifices that NHS staff on the front line have to make on a daily basis.

'You've certainly drawn the short straw today,' said my mum to him jovially.

'Trust me, nothing I could see on an A&E ward could be more traumatic than watching England at a World Cup,' he replied. Oh, how pessimistic adults can be, I remember thinking.

Twenty minutes later, I understood what he meant. For a young boy with a pretty messy injury, I'd entered A&E in a relatively calm state. Several hours later, I left balling my eyes out, having just seen David Batty hit a penalty that was on a par with George's first attempt in the corridor of The Royal Marsden. Thinking about it now, maybe I'd taught him a valuable lesson that day? That, as an Englishman, penalty shoot-outs would only bring him pain and suffering. I had probably done him a great service.

As the doctor had pointed out to us as we were discharged: 'The pain in your knee will get better, the pain of supporting England will not.'

Unfortunately, however amazing and awesome our NHS health professionals are, there is literally nothing that they can prescribe that can help with that!

SANDI TOKSVIG

I've had my life saved several times which, if there is a plan, suggests I ought to have some purpose for being here. The first time I was a small baby. I contracted tsetse fly fever, thanks to some oversharing fly, and my fever soared to alarming heights. We were living out in the African bush and, as the story goes, the flying doctor was called and arrived by small plane. I must have been boiling because apparently he declared, 'Put her in the fridge.' I've always been small and certainly as a baby I was tiny enough to go in beside the eggs. I don't think this is actually a recommended medical procedure, but here I still am. The upshot is a fondness for the fridge over and above it being a place to keep fruit jelly.

The second time medicine intervened to keep me on the planet was when I was twenty-one. I was at university, in my second year, and gradually had become less and less able to eat. I would consume a little and then be unable to force down any more. My fellow students loved sharing a meal with me because I always ordered well and left much, which they could then gorge on. I have never been interested in my looks. My wife says I am the only person she knows who doesn't look in a mirror to comb her hair. As a consequence, I had not noticed that my small appetite did not seem to marry up with an ever-expanding stomach.

During the summer holidays, I visited my maternal English grandmother, a forthright woman not awash with empathy.

'Are you pregnant?' she barked at me.

I shook my head. I was a secret lesbian, so pregnancy had not occurred to me as a worry, but I didn't want to say so. Granny took me to her local doctor, a woman I had known all my life from our annual English visits. Dr Johnson had seemed old for as long as I could remember. How many times I recalled her coming to my grandmother's house for one thing or another with her small black doctor's bag, which, in a Mary Poppins-like manner, always seemed to have the very thing in it that was needed. She examined me. My stomach was very large indeed and I was sent to the Middlesex Hospital in central London. It was 1979 and, unbeknown to me, they had something still very rare in the NHS – an ultrasound machine. I think everyone was still going with the pregnancy theory as I kept getting asked about it.

I remember the grainy pictures that flickered on a screen as they examined me. I appeared to have swallowed a rugby ball. This is not usual and caused some consternation. I think the machines are more refined now. There was talk of tumours and cancers; surgery was scheduled immediately and soon I had a scar which ran like a railway line from one side of my stomach to the other.

It turns out I had an ovarian cyst so large it was written up in the journal *The Lancet*, alongside other matters of medical interest. Seven pints of fluid in it. The surgeon joked that they had all had to wear wellingtons. This unwanted sac of fluid had pushed against all my organs and stopped me being able to eat. My stomach was large but in truth I had gone down to five and a half stone. I was starving.

The recovery took many weeks and I got to know Broderip ward at the Middlesex extremely well. Here the team of nurses were led in an old-fashioned way with Matron shooing doctors away when she thought they had had enough time with the patients. They were glorious women who managed a tricky mixed collection of patients, including an elderly woman with advanced dementia who drove us all mad waking us at night to see who we were. The nurses were kind and patient and dealt with my distress when I had a sort of post-traumatic realisation at how ill I had been. How little attention I had paid to my body. They healed every part of me. They changed bandages, they brought tea and, most of all, they made me laugh.

Hard to believe that an ultrasound machine was still cutting edge when it helped to keep me here, but the NHS has ever moved forward and adapted. In 1987, Broderip ward became the UK's first ward dedicated to caring for HIV patients, when it was opened by Princess Diana. It changed the face of how people were treated and brought dignity to the gay community. I was proud to have ever been in it.

There are photographs of the patients from that time taken by Gideon Mendel, which were published as a book called simply *The Ward*. In one of them, a nurse is leaning across her patient to give him a kiss on the cheek. It represents all the caring and kindness which I know first-hand from those who work in the NHS. My own daughter is now a doctor in their service and I wish Dr Johnson were still alive so I could tell her.

I still have that long scar which divides me in half. I've been thinking of having the fastening for a zipper tattooed at one end for no other reason than it would have made the Broderip nurses laugh. I thanked them then and with all my heart I thank them again.

The Middlesex Hospital is gone now. In its place stand expensive flats and offices. A hospital had stood on that street since 1757, but was closed down in 2005. Perhaps with hindsight one should still be standing there now.

PETER CAPALDI

When you answer a call from your teenage daughter, you really don't want to hear a man's voice (from the big medical tent at the Notting Hill Carnival) on the other end saying, 'Oh hello, this is the medic, I've got your daughter's phone . . .'

'?!'

'Don't worry, she's all right.'

Dehydrated, stressed, fainted (and not even from booze, either). But 'all right'. And brilliantly looked after by the NHS ambulance crew that got her to A&E – and had already thoroughly checked her over – by the time we, and our shredded nerves, arrived.

Babies and bunions. Strokes and heart surgery. The NHS is always there. Not always on time. Not always perfect. Not always as loved and appreciated as it should be. But constant.

Indeed, it is such a perennial part of our lives that I think lots of us don't realise (or have forgotten) that there was a time when it did not exist, and that there are places in the world where universal healthcare, available to all and not defined by the capacity to pay, is just a dream.

Like most people, my interactions with the NHS have been many and (thankfully) generally low-key. I was born in Stobhill Hospital in Glasgow (quite possibly while actor Richard Wilson was serving as a research assistant in the laboratory there). I had the usual childhood tonsils operation (there was a fad for it in the 1960s) and, while at art school, rock and roll gave me a tragi-comic visit to casualty one Christmas Eve after a gig with my band in Paisley. Sweat and the Atomic Gel holding up my exuberant quiff combined and by the end of the gig I had to be guided into a taxi destined for A&E and a soothing eye bath accompanied by some hair product 'advice'.

After he had the obligatory west of Scotland 1980s heart bypass, my dad was 'all wires and tubes' in intensive care, but gave us an encouraging thumbs up. He compared the operation to having been 'hit by a bus'. This didn't put off another freshly carved up and heavily tattooed patient (tattoos were not fashionable in those days) from stomping around shouting, 'I want out of this ****ing place!' A Glasgow hard man looking for trouble, he was put in his place by an old-school Glaswegian sister who told him to turn it down and stop disturbing the other patients.

In the 1990s, I was playing an eighteenth-century fop in a BBC costume drama (rouge, lipstick, a beauty spot and powdered bouffant) and one scene called for me to be confronted – and punched – by the squire. The stunt punch went well in rehearsals and on the other actor's shots, but when the camera reversed onto me, 'action' is the last word I remember before I woke up on the floor with everyone mysteriously looking at me. When I put my hand to my temple it was wet. My fingers were covered in blood.

I saw my co-star, the squire, sitting ashen-faced. Is it OK to say that the squire was played by Brian Blessed? And, also, have you seen the size of his fists? They look like Christmas hams. We had mistimed our movements and Brian, to his eternal regret, had knocked me out.

And then there were two ambulance men. One of them said with grim urgency, 'Get his wig off!' As his partner reached for my powdered locks, I squeaked out the words, 'It's my own hair!' (I have a *lot* of hair.)

Ambulance, casualty and nine stitches right along my eyebrow later ('opened up like a boxer's'), and still fully made up and wearing a gold jacket, jabot, laced shirt, breeches, white tights and ornate buckled shoes, I was told I was being kept in for observation due to a head injury.

And much as I love the BBC, no one showed up at the hospital except for the slight and nervous figure of my dresser. Could they have the very expensive costume back? 'Luckily' it hadn't been stained with blood, and the costume department wanted to make sure it was kept safe. He had a poly bag which contained my own trousers and shirt. But no shoes.

So, in a full ward, surrounded by the worse for wear and the unlucky, I spent the night in a hospital gown, fully made up and powered like Quentin Crisp on a bad day. The doctor, of course, was completely unfazed. He'd seen a lot worse.

My family were (and are) much prouder of our NHS connections than anyone having a foothold in show business. My cousin Senga became a doctor, which in itself was a big enough deal. But then she married Harry Burns, who not only became chief medical officer for Scotland, but a sir into the bargain – making Senga *Lady Senga*. And, of course, she was the resident go-to person for countless Capaldi relatives with any medical query. If Senga said it, it was gospel.

Being Doctor Who was one thing, but being a real doctor – that was on a whole other plane. Though when my mother was spending her last days in hospital in Glasgow, she loved to report that they called her 'Doctor Who's Mammy'. As my sister and I sat with our ailing mother on New Year's Eve, it was strangely comforting to hear the nurses behind the screens celebrating with Chinese takeaway, accompanied by the distant sound of revellers on the high street. The nurses' care and compassion as we were losing our mum, and the tact of the doctor, are hard to overpraise.

I was grateful, and I am grateful, to know that my family has been, and will be, cared for throughout our lives by probably the greatest idea that anyone has ever had: the NHS. Thank you.

EMILIA CLARKE

Why I Clap

I believe my first real memory of my beloved NHS is of the kindness of a nurse when, as a very sugar-dependent three-year-old, I'd climbed to the top of the bathroom cabinet and devoured an entire jar of sugary multivitamins under the cover of night-time. She very calmly told my slightly overwrought and worried mother that once the sugar high had worn off, there would be no lasting damage and that I would indeed be fit to fight another nursery day, and then some.

Then there was the doctor who stitched up my brother's knee when he thought jumping onto broken glass while unpacking boxes in our new home was the best idea. He got a badge for bravery that day.

The memories I will hold dearest, though, are ones that fill me with awe: of the nurses and doctors I knew by name when, in the weeks after my first brain haemorrhage, we watched the passing of time and the passing of patients in the Victor Horsley Ward at the National Hospital for

Neurology and Neurosurgery in Queens Square, London.

The nurse who suggested – after everyone else in A&E struggled to find an answer when I was first admitted – that maybe, just maybe I should have a brain scan. She saved my life.

The anaesthetist who miraculously kept me giggling along with my entire family as he talked me through the process of what was about to happen to my brain and then counted me down from ten.

The surgeon whose skill, quick thinking and sheer determination saved my life, while never letting on how close to death I had been.

The countless unthanked nurses who changed my catheter and cleaned up my bile-coloured vomit on the days when I couldn't even manage water. The nurses who washed my body with care, generosity and love when I couldn't walk or sit, who carefully put me in pyjamas I recognised as my own when my morale dipped below the surface, with as much kindness as if I had been their own daughter.

The cleaners who mopped the floor when my bedpan fell to the ground, shame and embarrassment filling the room along with disinfectant, and then a reassuring smile and a knowledge that they'd seen worse.

The doctors who talked to me as a fully functioning adult on their daily rounds, looked me in the eye and told me everything they knew, but always finished with a joke that was enough to assure me that I was actually in control of what was happening to my brain and body, despite all evidence to the contrary.

The phlebotomist who took my bloods every day who normally worked on the children's ward, but who was sent my way by the nurses after a day or two of trying to get a hold of my tiny hidden veins. Yes, I got a lolly. No, I didn't feel a thing.

The cooks who made my fish in white sauce with peas every day, despite it being a child's meal; it seemed to be the only thing I could eat at least some bites of. The nurse who brought my best friend's note up from the cold outside when he'd missed visiting hours but knew how much it would help me get through the night. She had been on her way home.

When I was in ICU following a severe bout of dehydration-led aphasia, during which I lost my ability to speak coherently, I heard the patient in the bed next to me in the final moments of his life. One of the incredible nurses on duty allowed my mum to stay next to me and hold my hand instead of leaving, as every other patient's loved ones were asked to do. She saw that, in this moment, she held my fragile mind, and its capacity to pray that I wouldn't be next, in her hands.

In all those moments, over those three weeks, I was not, not ever, truly alone.

These are only a handful of memories from those weeks as I waited to see if I could leave with my life intact. I could write a book about every moment I've spent in the loving warmth of our NHS. When my dad passed away, the humour, love, kindness and empathy I received while witnessing his final moments are among the greatest acts of humanity I have known, or will ever know.

And to each of us who has our own memories, moments – both light and low – they are just that. Moments in our lives. To the incredible people who work in the NHS it is their every day – their morning, noon and night – and we are made to feel as if each of ours is unique, treasured, one of a kind, when we all know that they have witnessed more deaths, births, heartbreaking tragedies and heart-soaring triumphs than any one of us could possibly fathom, and will continue to, every single day.

This beloved institution that we have as a nation, that is OURS – all of ours. These brave compassionate, kind, intelligent humans have held us up in every one of our darkest hours, so it seems only right that now, in these unprecedented times, the least we can do is hold them up during theirs.

So every Thursday I clap. And I will keep clapping, keep donating, keep seeking out ways to show my thanks, because, NHS, you've given me enough reasons to last a lifetime.

JAMIE OLIVER

What an incredible outpouring of love our great nation has shown for the NHS and key workers during this pandemic. It's absolutely deserved and a joy to witness. The weekly clap for carers is super-powerful and shows just how much love and respect there is for all those wonderful, committed, passionate people that are working so hard, every day, to keep our health service going.

Dear reader, if you want to do something to help your local NHS heroes, I think it would be really nice to rustle them up something delicious and nutritious to eat, so it's one less thing for them to worry about. I'm sharing here a couple of my classic recipes – a flavour-packed, versatile veggie chilli as well as some cute little rye bread scones, which come out of the oven ready-portioned. Cook up a batch of chilli, portion it up with the scones and drop it off at your nearest NHS establishment. You just know that's gonna be appreciated.

VERSATILE VEGGIE CHILLI

Hearty and delicious, this alternative to traditional chilli con carne is sure to go down a treat. Simply double or triple the recipe to make a bigger batch.

SERVES 4
TOTAL TIME 1 HOUR

500g sweet potatoes or butternut squash
1 level teaspoon cayenne pepper, plus extra for sprinkling
1 heaped teaspoon ground cumin, plus extra for sprinkling
1 level teaspoon ground cinnamon, plus extra for sprinkling
Olive oil
1 onion
2 mixed-colour peppers
2 cloves of garlic
1 bunch of fresh coriander (30g)
2 fresh mixed-colour chillies
2 × 400g tins of beans, such as kidney, chickpea, pinto, cannellini
2 × 400g tins of quality plum tomatoes
lime or lemon juice, or vinegar, to taste

METHOD

- Preheat the oven to 200°C / 400°F / gas 6.
- Peel the sweet potatoes (or peel and deseed the squash), chop into bite-sized chunks, then place on a baking tray.
- Sprinkle with a pinch each of cayenne, cumin, cinnamon, sea salt and black pepper, drizzle with oil

then toss to coat. Roast for 45 minutes to 1 hour, or until golden and tender.

- Peel and roughly chop the onion. Halve, deseed and roughly chop the peppers, then peel and finely slice the garlic.
- Pick the coriander leaves, finely chopping the stalks. Deseed and finely chop the chillies.
- Meanwhile, put two tablespoons of oil in a large pan over a medium-high heat, then add the onion, peppers and garlic, and cook for 5 minutes, stirring regularly.
- Add the coriander stalks, chillies and spices and cook for a further 5 to 10 minutes, or until softened and starting to caramelise, stirring occasionally.
- Add the beans, juice and all. Tip in the tomatoes, breaking them up with the back of a spoon, then stir well.
- Bring to the boil, then reduce the heat to medium-low and leave to tick away for 25 to 30 minutes, or until thickened and reduced – keep an eye on it, and add a splash of water to loosen if needed.
- Stir the roasted sweet potato or squash through the chilli with most of the coriander leaves, then taste and adjust the seasoning, if needed.
- Finish with a squeeze of lime or lemon juice or a swig of vinegar, to taste, then scatter over the remaining coriander. Delicious served with yoghurt or soured cream, guacamole and rice, or tortilla chips.

NUTRITION 369 calories, 10.1g fat (1.6g saturated), 21.5g protein, 58.3g carbs, 14.4g sugars, 0.9g salt, 12.9g fibre

To see the recipe image, please visit jamieoliver.com/veggiechilli

RYE BREAD SCONES

MAKES 24 SCONES
TOTAL TIME 40 MINUTES

Olive oil
500g natural yoghurt
250g stoneground rye flour
250g self-raising flour, plus extra for dusting
1 heaped teaspoon bicarbonate of soda
1 teaspoon runny honey
2 large free-range eggs
3 tablespoons organic jumbo oats

METHOD

- Preheat the oven to 190°C/375°F/gas 5. Grease a couple of large baking trays with oil.
- Tip the yoghurt into a large mixing bowl with the flours, bicarbonate of soda, honey, one egg and a large pinch of sea salt.

- Stir with a fork until everything just comes together, then get in there with your hands and shape the dough into a ball.
- Dust a clean surface with flour and flatten the dough into a large disc, roughly 2.5cm thick.
- Stamp out twenty-four rounds with a 5cm crinkle cutter, dipping the cutter in flour if the dough starts to stick, placing them on the oiled trays as you go.
- Beat the remaining egg and use it to eggwash the scones, then sprinkle over the oats. Bake in the centre of the oven for 20 to 25 minutes, or until puffed up and golden.

NUTRITION 98 calories, 1.8g fat (0.8g saturated), 3.4g protein, 18.1g carbs, 2g sugar, 0.3g salt, 1.9g fibre

To see the recipe image, please visit
jamieoliver.com/ryescones

NISH KUMAR

The last time I went to hospital, it was perhaps for the most pathetic reason possible. In February 2020, at the age of thirty-four years, I got the end of a cotton bud stuck in my ear. Now, I understand there will be two immediate points of criticism and I'd like to address them:

1. 'Cotton buds are so bad for the environment, how dare you contribute to the plastic waste. It's almost as if that Extinction Rebellion sticker on your laptop means nothing' – they were plastic-free ones.
2. 'Everyone knows you shouldn't be using those, they're bad for your ears' – admittedly I have no defence here. I know I shouldn't be using them but it feels extremely satisfying in the short term, even though it's ultimately self-destructive, like drinking alcohol or watching *Question Time*.

So, I was having a good old dig. But then I withdrew the magic stick, only to discover the cotton end had vanished.

In the words of the unfortunate teacher who marked my year nine geography coursework: 'This is dreadful and it's entirely your fault, Nish.'

I snapped into action and called the cotton bud a d**k h**d. I then got my girlfriend to stand on a chair and look into my ear, in yet another incident in the course of our relationship that she describes as 'a boundary violation' and 'a test of my saint-like patience'.

She couldn't see anything. We then consulted our in-house physician Dr Google, a wildly underqualified individual whose diagnoses are too broad to be of use – 'you are both absolutely fine and utterly doomed' – and whose prescriptions are either 'boil your whole head' or for you to use a product that enlarges a part of your body that is not in any way affected by your current predicament.

Frankly, the only thing to recommend Dr Google is that he operated in the same hospital as Dr Bing, whose only diagnosis is 'please just ask Dr Google – I have no idea'. And of course there's also the rather old-fashioned Dr Ask-Jeeves, who always just sends you to the apothecary to fetch leeches.

Anyway, with no useful word from Dr G, I called NHS 111 and they gravely informed me that I'd have to go to hospital. So at 1 a.m. I headed over to the hospital and we waited in A&E with people who had legitimate reasons for being there, whilst I was there because something I knew I wasn't supposed to be doing anyway had gone wrong.

The doctor who saw me would have been well within

her rights to smack me in the mouth. Instead, she patiently listened and then had a look in my ears with an otoscope (thank you, Dr Google).

In that situation, it turns out there are two key phrases you don't want to hear:

1. 'OK, there's nothing in there. I'm guessing it fell off after you pulled it out. The reason you are feeling pressure in your ear that you think is the cotton, is that you have a substantial build-up of wax and when you used the cotton bud you pushed it deeper inside.'
2. 'By the way – big fan.'

Perhaps the worst place to be recognised as any kind of public figure is when a doctor is looking in your ear, at a candle shop's worth of wax that has tricked you into thinking you had blocked your ear with the end of a cotton bud. I got the name of some ear drops and we skulked off with my girlfriend saying something about a 'last straw'. I couldn't really hear her because of the wax.

Anyway, why am I telling you this story? Well, a few reasons. Firstly, the doctor recognised me – so I am famous. That's important to note. There are a lot of high-profile people in this book and I feel insecure about being included, so I'm really just putting this out there: someone recognised me from television, OK? So that means I'm the same as Paul McCartney.

Second, at no point did it ever cross my mind that this could ever cost me money. I wasn't being charged despite it all being my fault. The NHS's motto should be 'free – even for morons'.

And third, because when I start to consider why I am grateful for the NHS, the simple truth is that I don't know where to start. How can I begin to express my gratitude to its founders, who pursued an ideal of a healthcare system that would be free at the point of delivery? A system born out of the rubble of the Second World War, as a country dragged to the brink dared to imagine a better future for itself and its citizens. Where do I start to express gratitude to the NHS staff present at my birth, at my grandfather's heart operation and my grandmother's pneumonia last winter?

How can I express my admiration for the dedicated healthcare workers, among them my cousin, who have endured a decade of budget cuts, and yet still, in the hour of our need, have disregarded their own safety to protect us from Covid-19?

The truth is, I can't. So I won't. And I'll just talk about my waxy ears.

KATIE PIPER

Dear NHS,

I arrived unconscious, partially clothed and faceless.

My identity melted away. My status wasn't important to you; neither was my profession nor my possessions. No judgement, just good-quality care and fair treatment for all.

Triaged as: 'Female, 24 years old, 22 per cent full thickness acid burns to face and body, ingestion of acid and currently no sight in either eye.'

I could have been just another admission to your already bursting wards, a number on your sheet. But you treated me as if I were your own. Like the daughter, sister and niece that I was.

It's fair to say we kind of moved in together, me and you, Katie and the NHS. You let me sleep over for three solid months. We started in ICU and I gradually tried out every bed in the house: high-dependency, burns unit, main ward, endoscopy, the eye clinic, ENT, physio, plastics, psychology, maxillofacial, orthodontic and dermatology.

Quite the list when we all thought upon arrival that the only additional room I would be seeing beyond ICU was the morgue.

I skipped a bit of our time together, spending some time in a coma, but that was when you turned your care to my traumatised family, supporting them mentally as if they were the ones sick in the bed. Answering the anxious questions with gentle but professional truths, sharing your years of experience to help us with our very new and raw experience.

Your compassion, empathy and dedication goes way beyond your job role. No lunch breaks, clock watching or legging it out of the door the minute the shift ends. Staying on past the end of shifts to hold hands, working unpaid overtime. Care and compassion is the driving force always at the forefront of your exhausted minds.

I didn't realise it on the day we met, but that was it, I would never fully leave you again; you have been in and out of my life, putting me back together and keeping me alive still to this day, twelve years on. We've been through some real highs and lows together, but you gave me a second chance at life when many feared my time was up. Then you helped me to give life to my two baby girls – from beginning to the end you held my hand once again.

NHS, you've taken me from victim to survivor, and finally to proud mother. How will I ever repay you?

I winced when I heard people publicly discuss you,

people who didn't understand you or what you were doing every single day; what you witnessed, what you absorbed and what you gave. As they debated whether we needed you, I felt an anger rise and a tear drop. You are the backbone of our country; take you away and we all fall down.

Thank you, NHS. I owe my life to you – you fixed me when I was broken, battered and hopeless.

You made the impossible possible and I will never forget each and every one of you. We are privileged, we are spoilt by you and you must not be taken advantage of. You are what puts Britain on the healthcare map. NHS, you truly are the heart and pride of this country.

Love,

Katie P

x

BRIDGET CHRISTIE

The NHS delivered both my babies. They were born in Homerton Hospital, Hackney, three and a half years apart. The first birth, in April 2007, wasn't as fun as I'd imagined. In fact, a doctor told my husband we were lucky. If it'd been fifty years earlier, they might've lost one of us. I don't know how old the doctor thought I was, but I remember being very insulted by that.

It didn't start well. I was two weeks overdue for a start, and when my waters finally broke, on my bedroom carpet, a mouse ran over them. We lived in a flat above a deli and were constantly infested. I screamed and shouted, 'Get off my amniotic fluid, you verminous twat!' It really ruined the moment. This precious liquid had kept my baby safe all these months and now a mouse had gone on it.

On the way to the hospital we got told off by the police for driving too slowly, and a bit further on we had to do a U-turn and were diverted because there was a body lying in the middle of the road in front of us. A lone policeman had just got there and was cordoning off the area with police tape. I thought about the man's own birth. He was someone's new baby once, a mother's pride and joy. He'd been held up, kissed and cuddled, shown off and photographed. He'd been fed and nurtured and read stories and tucked up in bed. And now he was here. In the road. Alone. I felt deeply and profoundly sad and we drove the rest of the way in silence. Then we couldn't find anywhere to park.

Unbelievably, I wasn't even IN labour yet and a midwife sent me home, which was annoying because we'd just paid for parking. I got home, lay on the sofa, had staring competitions with the mice and ate a hot curry but none of it helped. I just wasn't progressing. It was like being back at school. The next day we went back in. Three shifts of midwives and thirty-six hours later, my baby still wouldn't come out. He's still like this now; he's just swapped my uterus for his room.

By now, the baby's heart rate was all over the shop and I wasn't doing great either, so it was decided the best thing to do was to take me down to theatre and 'get this baby out'. I thought all labours were like this and wasn't overly concerned. On the way down, I kept vomiting into a cardboard bowler hat and had a consent form shoved in my face but didn't sign it. My husband put some weird blue plastic clothes on and someone put my feet in stirrups. Everyone was really jolly and made me feel very confident. After the maximum amount of goes with a ventouse and some forceps from medieval times, I was told it was really time for an emergency C-section now and could I just sign the bloody form?

I begged them to try one more time, which they did, and my son was finally born as 'Teenage Kicks' by The Undertones blared out of the radio. He was enormous, grey, had a cut over his eye from the forceps and looked really annoyed. The doctor stitched me up as a porter watched on with a reassuring aloofness. I probably should've had that C-section because that birth gave me massive haemorrhoids, which I've had ever since. Even my health visitor commented on them. I don't think she'd seen anything like it. My friends tell me I can have them surgically removed but I've chosen to wear them with pride, like a badge of honour.

My second birth was easy peasy. During my last trimester, I developed a rare liver disorder called obstetric cholestasis. It can be quite dangerous in pregnancy and in some cases cause stillbirth. I'd been maniacally scratching the soles of my feet and my palms like a madwoman for a couple of weeks. Then, by pure chance, I read about these weird symptoms in a pregnancy booklet called 'Emma's Diary' I'd picked up from my GP surgery (thanks the NHS!).

I called my GP and told her I might have this liver thing. The surgery was just about to close but she told me to come in, give her a urine sample and she'd send it off to Homerton. At 5 p.m. on 31 December 2010 she called me to say my results were back, I did have it and I should get down to the maternity ward and have the baby checked over, and that's where I spent New Year's Eve. Happy New Year, Homerton!

The baby was constantly monitored for another week and a half and then a nice consultant suggested I be induced, just to be on the safe side. It was such an easy, calm labour. I remember saying to my midwife, Cheryl, 'When is everyone else getting here?' and she said this WAS everyone and looked a bit insulted. I couldn't believe it! ONE person! My son's birth needed about twenty!

The baby was born very quickly. Too quickly for any proper pain relief. But it was fine. I knelt up, faced the wall and prayed, which helped a lot, even though I'm not particularly religious, and my precious, tiny baby was born arm first, like Superman. I'm a massive fan of pain relief and I would've absolutely had everything available to me if there'd been time, but this was a good pain and I'm glad I felt it. The only complication was that the umbilical cord was so short Cheryl had to call for someone else to cut it while she held the baby. Thanks to the NHS for my beautiful babies and my magnificent piles.

LORRAINE KELLY

It was February 2012. One minute I was nervously sitting atop a horse for a charity challenge and the next I was being swiftly, expertly and gently moved onto a stretcher and rushed to hospital by kind paramedics.

I had always been a bit scared of horses and this was only my second riding lesson. As it was all to raise funds for a children's charity, I thought I would be perfectly safe. I couldn't have been more wrong. The horse had reared up, thrown me off its back and smashed a massive iron-clad hoof into my thigh as I lay helpless on the ground.

I was in complete shock, lost three pints of blood and needed help urgently. And, of course, the NHS was there to pick up the shattered pieces.

In the ambulance I was given morphine, but much more importantly I was given reassurance. They held my hand and told me I was doing fine and not to worry that the blue light was flashing and the siren blaring.

I was rushed to the nearest hospital, St George's in

Tooting, London, but I honestly don't remember all that much about being taken to A&E. It was all a blur of busy but controlled efficiency.

The one thing I knew for certain was that I was in the best possible hands and so I wasn't scared. Though I was lucky. A few centimetres either way and the artery could have been severed or my pelvis completely shattered.

It is completely crazy what goes through your befuddled brain in times of crisis. As they cut through my clothes and underwear, I remember thinking that my mum was right (she's always right) and I was glad I had heeded her advice of wearing clean, matching knickers and bra, 'In case you have an accident.'

I was in London but my husband Steve and daughter Rosie were miles away in Dundee. As they headed south, my fantastic friend and colleague Emma Gormley rushed to the hospital. She says she will never forget seeing me being wheeled into the operating theatre and apologising for probably not being able to make it into work the next day.

The operation lasted five hours. The skill of NHS surgeon Martin Vesely and his team meant I didn't need a skin graft, but I had to have hundreds of stitches on a wound that looked like a massive shark bite.

I had to stay in hospital for over a week to recover and Martin was brilliant. He had the best possible bedside manner. Every morning on his rounds he would give me an update and tell me clearly how I was progressing, and then ask if I had any questions. Like most patients, especially those on strong painkillers and sleeping pills, I struggled a bit to take everything in. But Martin did an amazingly kind and considerate thing. He would go to the bottom of my bed and simply wait for a heartbeat, giving me just enough time to think to ask him something that had been preying on my mind. He also treated me like a person and not 'a serious leg injury' and that made all the difference in the world.

From the nurses to the cleaners, the physios to the volunteers who came round with books and magazines, I was so well looked after, but it was clear resources were stretched to breaking point.

One of the hospital wards assigned to the elderly – most of whom had some form of dementia – had to be closed due to an MRSA infection and the patients were moved in beside us. The patience of the nursing staff was just unbelievable. They had to deal with poor bewildered souls wailing in distress all through the night as well as aggressive behaviour that came out of nowhere. One nurse even had a full bedpan thrown at her, but she simply cleaned herself up, changed her uniform and carried on with her shift.

Last year I was able to go back to St George's to say a proper thank you to all of the staff at the coalface who helped put me back together again. It was so good to have a chance to express my gratitude face to face and I found it extremely emotional.

As you could expect, they all said they were just doing

their job and of course my accident, although it scarred my leg and my life, was just one tiny droplet in a vast ocean of routine emergencies they deal with on a daily basis.

All day, every day, they bring new life into the world, heroically battle death and devastation and give comfort to the desperate and the bereaved.

Then they get up and do it all over again.

This latest crisis has shown us all just how bloody lucky we are to have our brilliant band of NHS workers taking care of us all.

We must make sure we tell them how much they are valued and never, ever take them for granted. And we need to make sure they are properly looked after and protected so they can continue to do their jobs, which, as we know, are the most important in the world.

ANDREW MARR

Like almost everyone, I've had lots of NHS experiences, from possible cancers to the births of my children. But by far the most significant was having my major stroke seven years ago. It had happened overnight and I had woken up lying on the floor unable to get up. It was terrifying and yet, weirdly, from the moment I was strapped onto a stretcher and carried downstairs into an ambulance to Charing Cross Hospital in Hammersmith, I felt I was in completely safe hands and that somehow everything would be fine.

From the rush across south London, siren blaring, to waking up in bed a few days later after just dodging death, it would prove to be one heck of a journey. There were bad moments. An operation to clear a blood clot in my carotid artery failed. My family were gathered together and told I probably would not make it, and then again, later on, to say that if I did, my brain would be severely affected and I would probably spend the rest of my life in a wheelchair.

Even after I came round, quite disabled, there was a long haul back through rehabilitation before I could walk even a short distance on my own, or talk clearly. Again, that was scary.

But what I remember today is the openness and the friendliness of the doctors, nurses and therapists around me. The doctors were frank in their explanations and treated me as an adult. The nurses worked endlessly but somehow retained a wry, salty good humour. Two physiotherapists, one from Eastern Europe and one from Australia, went many extra miles to get me going – even rigging up a homemade autocue in the gym so that I could practise before getting back to work ... when my first job was to interview the then-prime minister, David Cameron. I felt I wasn't surrounded by sickness, but by good people.

From Charing Cross Hospital I went to the National Neurological Hospital at Queen's Square for further therapy on my non-functioning left arm. Again, just great people, a thoroughly democratic air in the wards and a huge deal of laughter. For a while, I had an agonising shoulder problem, eventually sorted with an injection into the joint. The jovial South African doctor, bearing a hypodermic syringe about the same size as the Second World War howitzer, told me: 'I've got good news, and less good news. The good news is that once I've done this, you will be fine. You will sleep like a baby. The not *quite* so good news, Andrew, is that this is going to *effing* well hurt.'

It did. But I quickly discovered that it's quite hard to howl and laugh at the same time. Our hospitals run on taxpayers' money, world-class training and inspiring dedication. But beyond all that, they run on humour, and humanity. It isn't the buildings or the awesome technology that we stand to applaud when we clap for the NHS. It's the grit, the realism and the gutsy humour of the extraordinary people who keep it going. I wasn't exactly lucky to have a stroke. But I was very lucky to have it here.

DAVID NICHOLLS

Extraordinary Machine

I was born and then, for years, nothing much.

I had jabs and check-ups of course, and dentistry – free in those days – and at fifteen my first pair of NHS spectacles, the silver-framed John Lennon specs I craved so much that I flunked the eye test to get them. At university there was an occasional GP visit; the cold self-diagnosed as glandular fever, the rashes and bad skin that came with forgoing fresh vegetables and sunlight. But there were no calls for an ambulance, no catastrophes.

Our two children were born in London hospitals and we took it for granted that there would be tests and scans, letters and leaflets and someone to answer the phone when the moment came. I'm not sure any birth can be described as straightforward, least of all by the father, but both children were healthy and for the next ten years, our encounters with the NHS were common-or-garden rashes, sore throats, a stitch in the scalp for my son when a trampoline party got out of hand. Each time we thanked the courteous, efficient staff and went back to our normal lives. Like electricity or water in the tap, healthcare would always be there, free at the point of delivery, based on clinical need and not the ability to pay. My father died in a hospital, cared for by compassionate staff. Beginnings and endings – perhaps that was all we'd ever need from the NHS. Perhaps there would be no disasters.

We were in Amsterdam on a half-term family trip when we noticed that my daughter was quieter than usual, not herself. Relaxation was discouraged on this holiday and so perhaps she was just exhausted by all that enforced sightseeing, the stomping from museum to landmark to museum. But on returning home, she remained pale and listless, pushing away her food, sleeping too much, the whites of her eyes the colour of butter. We took her to the GP, expecting that there'd be a few days off school. Some blood was taken.

And then, in a lab somewhere, a phone call was made and somehow it was as if some extraordinary machine had come into life. We were told to go to hospital now, immediately, to pack a bag, she might be staying. On the ward, there were more tests, a cannula was fitted, large doses of antibiotics, vitamins and steroids administered. There was no diagnosis but we were left in no doubt that this was serious and would require more than a day off school.

Questions. Had we come back from abroad? India? No, Holland. Eaten any seafood or raw meat? Gone swimming in rivers? Scans showed swelling and scars on her liver, the word 'cirrhosis' was used, surely a mistake because that's what alcoholics got and this was a ten-year-old girl. Throughout the injections and ultrasounds and consultations, she was patient, polite and upbeat but every now and then a wet glint in her eye betrayed her fear. A harrowing biopsy and then the diagnosis: auto-immune hepatitis, the body attacking itself. We celebrated her eleventh birthday on the paediatric liver ward.

She remained in hospital for the best part of two months, my partner and I taking turns to stay over, sleeping uneasily on the pull-out bed by her side, fidgeting through the perpetual jet lag of hospital life. During the long days, we'd watch the young patients come and go, many in a far more precarious state than our daughter, the parents fraught and haggard. We learned the nurses' names and acclimatised to the rhythm of the ward, saw the long shifts begin and end and came to understand that the events which had seemed so terrifying and life-altering to us were the daily business of here. The machine had not come into life. It had been running all the time, and it was the quiet expertise, patience and dedication of the staff that kept it humming away.

Now, in this current crisis, it is operating at its highest pitch and with the additional dark twist that those who work to save the lives of others are putting themselves at risk in doing so. God knows we owed a debt already, but every time I turn on the news these days I'm reminded of the limitations of the words 'thank you'.

Some day the crisis will pass, albeit with a terrible toll, and the NHS will return to the everyday task of keeping 70 million people alive and well. In the meantime, as I write, my daughter is upstairs in her bedroom, learning the endings of French verbs. She is fine, happy and healthy thanks to the workings of that extraordinary organisation – something for which we will always be grateful, will never take for granted.

KONNIE HUQ

My parents came over to the UK from Bangladesh in the 1960s after my dad graduated and was offered a job with Prudential in Holborn. Although it was heart-wrenching for them to leave their family and friends behind, they wanted a better life for their children than they felt their massively overpopulated and impoverished country could ever offer. After a few years spent settling in, they started their family – my two sisters and I were all born in NHS hospitals and so our cradle-to-grave journeys began.

My mum, although only basically educated, was an extremely intelligent and incredibly resourceful woman. She was also a whizz in the kitchen and could conjure up feasts out of nowhere. Even when she'd been working full time and fasting for Ramadan, I remember she would come home and rustle up a tableful of dishes for *iftar* (breaking fast) – every single one of them delicious, even though she hadn't been able to try a thing herself.

My earliest memories of my mum are of her comforting me when I was ill with a fever or a tummy ache. The calming touch of a mum by your bedside stroking your head is irreplaceable. If the illness persisted the next day and I had to stay off school, we'd go to the local doctor's surgery. They were always on hand to give us the magic green paper. She would swap it for medicine; sometimes it was even banana-flavoured. Taking it made me feel grown-up. The whole process was reassuring. I felt safe. I'm not sure at what point I actually realised how brilliant the magic green paper really was, that it let ALL children – no matter who, and many more besides – have free medicine to make them better if ever they got ill. Was there ever a penny-dropping moment or was it something that I just took for granted?

When I was fourteen, my parents saved up enough money to take us all back to their homeland. Bangladesh is a beautiful country, but one with 161 million people living in a space just over half the size of the UK. I remember my first encounter of the capital, seeing people hobbling about with missing limbs, blind people begging, poverty and illness everywhere. No magic green papers here.

Growing up, I was lucky enough never to need much more than the little green papers until my thirties. Though admittedly I did slip and hit my head in the *Blue Peter* garden one time! Thankfully, I was given the all-clear one quick CT scan later – in an NHS hospital, of course.

All that changed when my dad got cancer. My smart,

jovial, good-humoured father. I was devastated. The suppliers of the little green papers came good again, this time supplying radiotherapy. Radiotherapy that prolonged his life long enough to see my children – both born in NHS hospitals. He died the same year I had my youngest. Much like my son had come into the world, my dad had gone out of it, cared for and looked after by amazing staff in an NHS hospital. I was by his side. They helped me too.

After over fifty years of marriage my mum was going solo. She had looked after Dad in his final years: cleaning him, feeding him, watching him deteriorate. Time for new beginnings. In 2012, when the Olympics came to London amid much excitement, my sister took my mum as a treat. But my mum found it overwhelming and claustrophobic. She wanted to leave instantly. Strange. She was always the life and soul, this was *her* kind of thing. We thought she had depression. But slowly other strange things started to happen. She began burning food. My mum had never burnt a single thing in all her years – for her to even overcook a dish was unheard of. Then more things changed. She began to look a bit scruffier – my mum, who had always prided herself on her appearance. She started to get forgetful – my mum, multi-tasker extraordinaire. She became more bad-tempered – my mum, usually so measured and good-natured. The suppliers of the little green papers did some tests. My mum had early onset dementia.

It progressed fast. Only a year on from the Olympics, we couldn't even let her out of the house alone. As I watched my baby and toddler learn new words, I watched her forget them. Every day, as their little brains absorbed everything around them, things fell out of hers at an ever-increasing pace. As my eldest ditched nappies, my mum donned them. Each time I saw her, I would mourn different aspects of her. *We'll never laugh about that joke again. She can't tell me that story any more. She won't be able to teach me that recipe now. Does she even know who I am?* It wasn't long before she was hospital-bound – the same NHS hospital my father had died in.

My mother, conjuror of amazing food, was losing the ability to swallow. What now? A feeding tube? Starve to death? Starvation for my mum who was always so intent on feeding others? The irony. So many questions. So many decisions. So many unknowns. The hospital staff – overworked and underpaid – guided us through every step of the way. The doctors, the nurses, the support workers – we all had our favourites. I felt my mum did too. I'd sometimes see a flicker of recognition in her face or joy in her eyes, but it was hard to know for sure.

Among the decisions the staff had presented us with was permission to not resuscitate. It was a sunny day in the hospital the day she passed away. As with my father before her, I was there, holding her hand when her eyes finally glazed over. The staff had known and prepared me for what was coming as though they were oracles of the future.

It brings me great comfort to know that when I go,

should I need them, the suppliers of the little green papers will be there for me too, just as they were there for my parents. They are here for us all. They won't leave us limbless, hobbling in the street. For none of us are immune and we'll all go in the end. I have a one-in-three chance of having the hereditary form of dementia my mum had. My aunt had it, my great uncle had it ... I'm one of three sisters. The NHS offered me a test to find out but I think I'll take my chances.

So, thank you NHS for caring for us and supporting us. Unconditionally. From cradle to grave. How amazing is that? How lucky are we?

KATHY BURKE

A pain
A worry
A chat
A bed
A wipe
A jab
A wince
A tear
A relief
A breath
A smile
A cuppa
And a thanks
To the NHS

SIR DAVID JASON

The most urgently I have needed the NHS was when I had an argument with a hover mower. I was renovating a cottage in Crowborough at the time and wanted to get the place looking nice for visitors. Reader, the advice never to mow the lawn in slippy shoes should be heeded, as mowing down an incline I slipped and my natural reaction was to pull whatever I was holding onto towards me, which, at that moment in time, was a whirring blade. Ouch! I found myself in an ambulance on the way to the local hospital.

The A&E department were fantastic and also skilled in cutting off my shoe, which I hadn't dared to remove myself in case I found a few too many loose toes in there. When a brave nurse eventually took a closer look, she found the big toe attached to my foot by the smallest amount of skin and two other toes badly injured (I hope you've had your lunch).

A surgeon was called and I ended up on an operating

table, followed by three days in hospital. When they told me off for not wearing the appropriate footwear for mowing, I very nearly didn't have a leg to stand on (well, a foot). All joking aside, the NHS were there when I needed them and took care of me in the most professional and caring way. Fortunately I wasn't working at the time, but throughout my career I have had to stand on my own two feet and, thanks to the NHS, I have been able to do so.

Footnote: wear sensible shoes when mowing. Three cheers for the NHS!

JUNO DAWSON

I am often quite embarrassed to be British. I want to remain positive, so I won't go into why, but I sometimes will think of things I like about being British.

1. The NHS.
2. Yorkshire puddings.
3. Our sense of the absurd; I truly believe no other nation in the world could have come up with *The Rocky Horror Show*, *Absolutely Fabulous* or *Fleabag*.
4. We make the best pop music. This is indisputable.
5. Sarcasm.
6. Our cavalier attitude to alcohol consumption (and how it horrifies Americans).
7. We have all the different weathers.
8. We swear LOADS and have the most colourful curse words. Again, it thoroughly *appals* the Yanks.

9. The Spice Girls.
10. The vast majority of people employ a robust, inclusive and tolerant 'live and let live' approach in their communities. They have open hearts and minds and treat people with politeness, at worst, and go so far as kindness on a good day.

But I'd like to return to the first one. The NHS is the best thing about the United Kingdom. Anyone who says otherwise is wrong, sorry. People all over the world point to our National Health Service as best practice. Politicians (Trump) who don't want to provide free healthcare for nefarious economic purposes are terrified of their citizens coming here and seeing what we have and demanding it back home.

The NHS isn't perfect by any stretch, but that's on a practical level because – spoiler alert – some politicians would rather strip it back since running something free is phenomenally expensive. The very richest people in the country often complain that they'd like to keep their money or buy golden helicopters instead of paying for the NHS via their taxes. Those people are the pits.

Anyway, rant over. The NHS is brilliant. It has done brilliant things for everyone I know. During the coronavirus pandemic, my father became very sick. We don't know what caused it, but he fell into a diabetic coma. Although the NHS was already stretched to breaking point, paramedics came to his house and saved his life. No, really.

After he was admitted to his local hospital, and feeling a bit better, the consultant told him he had been mere hours away from kidney failure or death. Even during the worst crisis in the service's history, those doctors, nurses and paramedics saved my father. I get my dad for longer because of those people, and they did not charge him once he was discharged.

Like most of us, I started my life in an NHS hospital – Bradford Royal Infirmary. Since then, they have patched up my head when I fell against a radiator as a kid; they took out my janky wisdom teeth; fixed a broken nose; gave me medication to aid my panic attacks and, yes, supported me in becoming a woman.

For me, becoming a woman required more medical intervention than it does for most. I'm sure there are some people who think my medical gender transition was a zany use of NHS resources, but I take it to be evidence that we have a system that *cares* about *people*. Whole people.

Each of us will need the NHS for different reasons. We all play hard and fast with all sorts of things that are bad for us: sunshine, booze, cake, bacon, fags, casual sex, Instagram. I think we should never question *why* people need the NHS because we *all* will. I needed the NHS for my gender.

I was deeply unhappy pre-transition. I could not find peace. It manifested in a smorgasbord of mental health issues and risky, ill-advised behaviours. In July 2014, I went to my NHS GP ready to be laughed out of his office. I wasn't. Instead, he listened, for half an hour and in great detail, to my story of how I'd come to this conclusion over the course of thirty years.

My GP referred me to a specialist clinic in Northampton where a further two doctors agreed I had something called gender dysphoria and told me they could offer me medical assistance if I wanted it. I did, and the rest is history. I take two pills a day, a nurse gives me an injection every three months and I pop back to Northampton once a year so they can check how I'm getting on. Needless to say, I'm happier than I've ever been. Transition was the medicine I needed.

So when I say, 'I wouldn't be who I am without the NHS,' I really, *really* mean it, on a literal level. They enabled me to *live*. They saved my life.

JILLY COOPER

I'm mad about the National Health Service.

For a start, if I lived in most other countries, as an eighty-three-year-old I would be bankrupted by forking out for the Everest of painkillers, sleeping pills, soluble aspirins, blood pressure tablets, statins, eyedrops, etc, that I need to take to get me through each day!

But apart from my free meds, I love the NHS because we have a miraculous local surgery whose charming receptionists always manage to find a way to fit one in with an appointment. Glamorous framed photographs of former doctors adorn the sky-blue walls of the waiting room and one only has time for a quick flip through *Country Life* or *The Lady* before a smiling doctor or nurse summons you, immediately making you feel better. One particular doctor in fact is so good-looking that ladies need pills to reduce rocketing blood pressure before they even go into his consulting room.

We are so lucky too here in the Cotswolds to

have marvellous hospitals with really sympathetic and understanding doctors, nurses and ambulance drivers, to name but a few. Gloucestershire Royal Hospital, for example, cared for me beautifully when, years ago, I was admitted for eight days after attempting to rescue a ladybird in my hall and, upon trying to set it free on the lawn, I managed to fall over some rocks and cut open my leg, which became badly infected. Nor did the same dear hospital ever reprove me a few years later when, after a much too liquid lunch, I sauntered down the garden to feed the fish and managed to crack four ribs and puncture a lung by falling into the pond. Their kindness, care and attention remain firmly in my memory.

Nor have I ever appreciated the NHS more than when my dear husband Leo was dying of that most creepy and insidious illness, Parkinson's disease. Not just our local doctors, but Lizzie, the darling district nurse, would drop in to brighten our lives and the National Health carers would drive through the darkest nights to help make Leo comfortable in bed. On the night he died, carers Hazel and Jen came and laid out his body, and have remained firm friends ever since.

So today, I am not surprised that National Health doctors, nurses, carers and support staff are risking their infinitely precious lives going into the valley of death on twelve-hour shifts with often inadequate protection. As brave as the Battle of Britain pilots, they are fighting World War III against this horrendous coronavirus.

One wants to clap and cheer them – not just at eight o'clock on a Thursday, but at every second of every hour of every day.

NHS stands for Nicest, Helpfulest Saviours. Please God save all you can.

EMILY MAITLIS

The thing on Milo's foot is getting bigger. No one knows where it came from. Or what it is. We have tried ignoring it, cajoling it, prodding it and staring it down. He returns from a school trip – eight hours walking a day – and now it is huge, angry and ugly. It looks like a hostile takeover of the whole big toe.

It needs to be seen.

At the Chelsea and Westminster Hospital it is greeted with something close to pathological ecstasy. They have not been shown a thing like this before. A thing asking to be popped and drained. The very mention makes Milo go white.

'This is what I became a doctor for,' the medic tells me, with the excitement of a kid who's just won a day out at Alton Towers. It is almost certainly not true, but I am loving his enthusiasm. The nervous fourteen-year-old at my side, meanwhile, is burying his fears in sulky teenage nonchalance. He is inscrutable. He cannot meet their gaze. He gives nothing away. And somehow the doctors manage to find this endearing too. Milo has the lowest pain threshold of anyone I know. He was born early, which probably has no bearing on anything whatsoever, but it's what I tell myself when I see him clench up with anguish. We have shouted down entire GP surgeries at the prospect of a flu jab. The vaccination needle has machete-like proportions in Milo's mind.

This time, it is inconceivable a doctor will take any instrument to his foot. I look between the doctor and Milo. I look between Milo and the doctor. There is a quiet stalemate where we all ponder, in slow motion, what could happen next.

Then, into the growing silence, someone mentions laughing gas. I forget if it was the doctor himself or the bemused nurse at his side, watching her young patient's stern, unyielding face.

Within moments, a cylinder is rolled into the room. It looks big enough to inflate 500 helium balloons. Milo's interest is piqued. This is a story he can take back to his friends. He is now genuinely curious to see what it can do.

They place a scuba-like breathing apparatus over his face. All I can see of it is furrowed eyebrows and sceptical eyes. He breathes. The furrow goes. And the jaw relaxes. He reaches for my hand. Yes, my hand! For the first time in a decade he wants to hold my hand! I refuse to believe it is drug-related. He is just feeling particularly tactile right now.

Then the first bubble of laughter washes over him. And the next. He is now holding my hand more firmly and gently swinging my arm, as if we are off for a summery stroll in the park. He's almost ready for a full-on hug. There are little sighs and gurgles. More giggles. This child is having the time of his life. I want some.

The doctor has meanwhile set to work on the foot thing. He works with a speed and intensity that suggests there is an optimum length of time for laughing gas inhalation. Too much and the patient can end up turning a bit green. Milo and I have entered a new land: we are Dorothy finding Oz, we are linking arms and trotting along with lions and scarecrows and poppies and lollipops.

How we laugh! The anecdotes we tell! The memories that surface! What a time to be alive!

I am brought back to the room by the sound of a rather surprising pop followed by an actual eruption. There is now essence of toe on the doctor's scrubs. And a look of utter satisfaction on the doctor's face. Milo has felt nothing. Giggling. Reminiscing. Hugging. In his own little laughing gas world. And within minutes it is all sewn back up. Stitches, sterilisation, bandages. The neatest package of toe I have ever seen.

'Milo,' I say tentatively, 'it's all done. This brilliant team have done it. You're done.'

But Milo is not quite ready to give up the gas yet. It has to be gently but emphatically prised from his tight grip. There is no hurry, the team tell us. Take your time.

But Milo is in danger of taking this too literally. He seems happy to settle in for the rest of the afternoon. When he realises the cannister is on wheels he suggests we perhaps take it home.

Finally, it is me, comically, who boots my own son from the hospital bed where he has managed, against all the odds, to have the time of his life during the most skilful, professional, efficient piece of outpatient surgery I can ever imagine.

I am trying to find our team to say thanks. It has been bewilderingly fast, effortlessly empathetic and highly impressive. But they have already moved on seamlessly to the next kid with a dodgy something else.

'Come back for a check-up on Tuesday,' we are told at the outpatients' desk. 'We can change the dressing with laughing gas again, if you need it.'

And at this, Milo instantly perks up. We'll be back.

Like Arnie, essentially. But with a slightly lower pain threshold.

BILL BRYSON

In the late summer of 1973, quite unexpectedly, I stumbled into a job as a nursing assistant at an old and magnificent psychiatric hospital in Surrey called Holloway Sanatorium.

I didn't know the first thing about nursing care. I was just a young American college student hitchhiking around Europe, and was supposed to fly home in a couple of days to resume my studies, but two girls I knew who worked at the hospital urged me, during an awfully agreeable evening in a local pub, to apply for a job. The hospital, they explained, was perpetually desperate for menial staff, so impulsively the following day I applied, and the next thing I knew I'd been given a big set of keys, two grey suits, some white lab coats and instructions to present myself at a place called Tuke Ward at 7 a.m. the following morning.

Tuke Ward was high up in the building with splendid views over the hospital grounds and village of Virginia Water beyond. The patients, all male and all long-stay, were placid and cheerful and more or less looked after themselves. They went off every morning after breakfast to occupational therapy or gardening detail and didn't return till teatime. The charge nurse, an amiable fellow aptly named Jolly, likewise cleared off just after breakfast the first morning and I seldom saw him again.

Having expected to be on the way back to the United States, instead I found myself in sole charge of an empty ward in a large English hospital. I passed the days sitting with my feet up, reading old copies of *Titbits* and *Reader's Digest* that I found in the back of a large store cupboard, and from these I learned all about this new and remarkable country that I had now become part of.

I have seldom been more enthralled. I learned that there was something you could eat in Britain called blancmange, a pastime called morris dancing, a drink called barley water. I learned of the existence of Morecambe and Wise, seaside rock, Belisha beacons, milk floats, Poppy Day and a kind of strange voluntary prison known as a holiday camp. What an intriguing country! Every page was a revelation.

Then a second wonderfully unexpected thing happened. A few days after I started work, I found a note from Mr Jolly instructing me to go to a nearby ward to borrow a bottle of Thorazine, a medication. So I went to the neighbouring ward and, while I stood waiting for the bottle to be fetched, I saw across the room a pretty young nurse sitting with an elderly patient, spooning food into his mouth and dabbing his lips with a serviette, and I remember dreamily thinking: 'That's just the sort of person I need.'

By chance, later that evening I met the young nurse at a social gathering, and we got to talking and it turned out that she *was* just the sort of person I need. Her name was Cynthia. We were married two years later.

So the National Health Service has given me a wife, a new country and nearly half a century of kindly, world-class healthcare. That's why I stand on the front steps on Thursday evenings these days and bang a pot with a wooden spoon and shout, 'Thank you, NHS!' at the top of my voice. I am so pleased and grateful that I have decided to keep up the practice for ever.

REVEREND RICHARD COLES

I used to have a parishioner – I'll call him Pete – whose life was chaotic, sometimes spectacularly chaotic, thanks to the extravagant cocktail of alcohol and psychoactive drugs he used to make his existence bearable, if not always easy. From time to time, this would lead to hospitalisation and, because he liked me, I would accompany him to A&E, each time longing that the wait would be short, the assessment straightforward, the admission unproblematic. I liked him too – his honesty, his comedy, his paradoxical self-possession through self-destruction – but he was really hard work sometimes and I was impatient with him too often, and too quickly.

One night, he was even more than usually distressed and I found myself sitting with him in a bay in A&E waiting for a doctor as he got crazier and crazier, tearing the cannula from his arm, raving, playing with the oxygen and then running for the exit so he could have a fag. Eventually, in the small hours, we got him admitted to a ward and I sat with him to see if he would settle, though I was so tired and so fed up with his relentless delusional commentary, all the more exhausting for the moments of surprising insight that were characteristic of him in extreme distress. Never a dull moment with Pete; but he was more than I could handle that night, and when a young doctor arrived I thought he might just sedate him and, duty done, I could go home.

But he did not do that. Instead, he asked Pete how he was – not an unusual question from a doctor to a patient – but there was something in the way he said it, something about him, that halted the flow..

Pete was silent for a moment and then said, 'Are you a man of faith?'

'Yes,' said the doctor.

He was of Asian heritage and had a beard, and from his name I assumed he was a Muslim. Pete could be less than fully inclusive in his sympathies and I braced myself for a tirade, but it did not come. He opened his mouth to speak but the young doctor just looked at him so steadily, so calmly, with such frank sympathy, that he said nothing. They looked at each other in silence for a while and then they started to talk, really talk, and before long the doctor was sitting cross-legged on the bed, facing Pete, who was suddenly calm and eloquent and gradually at peace. He fell asleep.

The doctor wished me well and went on to his next patient. I thanked him, but I thought inadequately, because

I knew I had witnessed something quite extraordinary, and – in all honesty – felt some shame that the doctor had found a way into Pete's distress that I had not. The weirdest thing was that, as I walked through the car park to go home, I tried to recall what the doctor had said but could not remember a single word. The moment I tried to summon it from my memory it disappeared.

I will support the NHS with everything I've got, because it exists to provide medical care for all of us, as best it can; because of the skill and professionalism of those who work in it; because it needs protecting from people who think it a cost-occasioning indulgence (at least until they need it). But I love the NHS because of what that doctor did for my parishioner on a dark night, in a hard-pressed hospital, in an undistinguished town, where miraculous care unexpectedly happened.

SIR TREVOR McDONALD

The brilliance of the NHS response to the ravages of the coronavirus reminds me of the reverence I've always had for doctors and healthcare workers who dedicate their lives to caring for us when we need them most. The memory of that reverence comes back to me whenever I manage, almost against my will, to tear myself away from the unbearable urgency and pain of the blow-by-blow accounts of how the virus has brought misery to hundreds of thousands of families, ripped communities asunder and sent the world into lockdown. In that depressing gloom, one light shines as brightly as the morning star – and that light is the selfless commitment and courage of workers in the NHS.

That thought takes me back to the island in the West Indies where I was born, because of the great respect we had for doctors and the medical profession in general. Of course, there was no NHS in Trinidad, but health and a desire to keep one step ahead of any passing epidemic were our major preoccupations. They consumed the lives of our parents. The place of doctors in our tiny island communities framed much of our thinking and our aspirations. It's no exaggeration to say that, in a way, we were encouraged to do well at school so that we could become doctors.

At my secondary school, when the time came to make choices about which subjects we wished to take into the sixth form, Latin was *de rigueur*. At the slightest hint of protest, it was pointed out sharply that a career in medicine was impossible if one avoided doing Latin. We didn't ask to be doctors, it was assumed that we should be, so I plunged reluctantly into translations of Virgil because I had no choice. That medicine was so prominent in the thoughts of tutors and parents was no surprise. We survived in comparative poverty and well-balanced diets were uncommon. Few general bouts of ill health passed us by without leaving their mark – this enhanced the absolute status of the medical profession.

'You should become a doctor,' was the constant refrain of parents and every neighbour who saw you carrying a schoolbook – without much success in my case, I must add, though it rang through my years at school.

Then I came to London and fell into the protective bosom of the NHS. My first doctor in south-west London demonstrated such a degree of empathy that I usually left his surgery believing that he must have, at some time, suffered an illness exactly like mine. He made me feel my ailments were his too. His successor was so generous in treating my occasional high blood pressure that she never failed to remind me that 'white-coat syndrome' was always a factor when a patient underwent the test. She was indulgent. She had a second home in Arizona and, partly as a way of deflecting attention from my medical problems, we talked American politics and about the sheriff in the state who acquired international notoriety for his tough regime on immigrants and prisoners.

I know no boastful or overbearing doctors. The front-office staff at my west London NHS GP's were unfailingly courteous. The nurses were the same. When I turned up for my annual flu jab, before I could utter a cowardly word about how squeamish I've always been at having needles stuck into my arm, I would be told it was already well known in the surgery and that it would be done as painlessly as possible. I've always felt guilty about the fact that I don't ever remember feeling any pain at all. That's why, when I think of our NHS in the context of the monumental tragedy of this virus, and this epidemic, I'm reminded not only of the care and concern shown by my west London health centre, but of the doctors and nurses who came out of retirement to do what they've always done: to help the sick.

I shake with anger when I hear stories about health workers forced to fashion protective gear from bin liners and old curtains. I find it much too painful to listen to the tales of those who've lost their lives. They've gone back into the NHS in the service of a cause; they were well aware of the consequences – surely the ultimate sacrifice. Sometimes, in the course of our lives, we're granted the privilege of hearing of the deeds of our fellow men and women who represent the finest quality in all humanity; the very best that they could ever be in any of us. That is the dazzling image I have of the NHS. May they be always afforded the encouragement to serve our people, whenever that service is most needed.

GREG JAMES

I am honoured to be asked to contribute to this collection as, quite simply, I wouldn't exist without the NHS. My mum had a problematic pregnancy as antibodies had developed in her blood, making it totally incompatible with mine. At thirty-seven weeks, I was induced and was in a critical condition. We owe the NHS everything. We wouldn't be a family without it. To all the staff who worked at Lewisham Hospital in 1985, thank you.

For obvious reasons, I was unaware of the full story, so I called my mum to ask her for a few details to write up for this book. What came back from her made me cry. Like, a real 'close the laptop and call her back' cry. I hadn't quite realised what she went through, and it became clear that her first-hand account is what you should be reading. So, Mum, it's over to you . . .

Dear NHS,

I have rhesus negative blood, something which I

was unaware of until I was first pregnant in my early twenties. Subsequent pregnancies became increasingly problematic as antibodies had accumulated in my blood. These antibodies were effectively hindering the healthy development of my baby in the womb. This made being pregnant worrying, to say the least.

Early in the pregnancy, I had a period of enforced bed rest with my GP visiting me at home. As things progressed, I had to visit Lewisham Hospital twice a week for plasmapheresis treatment, which involved a needle in one arm to extract blood, which was spun to remove harmful antibodies and then returned via the other arm. It was painless and I got used to travelling between Bromley and Lewisham on the 208 bus. Though at the first treatment, my husband almost fainted at the sight of the needles. Your staff were always amazing and made it such a pleasant social occasion, with the customary tea and biscuits, of course.

In order to monitor the condition of the baby throughout the pregnancy, I had regular amniocentesis, which involved the insertion of a very long needle into my tummy to draw off fluid. I still have the puncture marks! Throughout this initial period, all the staff at Lewisham Hospital, including the gynaecologist, were fantastic, always very reassuring, competent and caring. I don't recall ever feeling unduly anxious during what could have been an incredibly lonely and terrifying time. I couldn't have felt more supported; they made me feel special and that gave me the strength to get through.

It was decided that Greg would be induced at thirty-seven weeks, which happened to be Greg's sister's birthday. She already hated the idea of a brilliant, sweet, handsome little brother (Greg told me to put that last bit in), so your good-humoured doctors agreed that we'd wait until the following week. The labour itself was stress-free, quite short and managed by an amazingly calm midwife, who presented me with my beautiful baby boy complete with a mop of thick, dark hair. There was even time for a quick cuddle before he was taken away to an incubator in the Special Care Baby Unit.

Owing to the increasing severity of Greg's jaundice, he was given a complete blood transfusion, closely followed by two more. This was by far the most anxious time because Greg was now an actual little person. The months of tests and procedures had become part of a routine, but now we were involved emotionally with our new baby. I remember that each visit to the Special Care Baby Unit was incredibly traumatic as we were never sure how our baby's health would be. However, as always, we were wonderfully supported by everyone who was looking after him. Miraculously, Greg's condition began to stabilise, so it was decided to give him a final top-up transfusion.

A few days later it was Christmas Eve and I went up to the unit to feed Greg. It was magical and I'll never forget it. What I saw typified the selfless, kind acts of humanity that makes the NHS so special. The dedicated team of special care nurses were quietly caring for the babies, fairy lights were twinkling and carols were playing softly in the background. Each baby had been given a small gift, left at the foot of its cot. It was a beautiful moment, so peaceful and uplifting and so emotionally charged.

I was joined by Greg's dad, who had been sleeping in the car park most of the week, unable to go home until he knew our baby was going to be all right. Thanks to all the wonderful NHS staff he was, and on Christmas morning we were allowed to go home with our little Christmas angel. As a family we will always be very grateful for all the love and care you gave us during that period in our lives.

Thank you, NHS.

Rosemary (Greggy's mummy) x

P.S. He hates me calling him Greggy in public.

MARY BEARD

I cannot claim that it was my very *best* Christmas, but it was one of the most warmly memorable. It must have been 1961 or '62, which would have made me six or seven. I was in the Royal Hospital, Wolverhampton, to have some complicated but decidedly unglamorous engineering work performed inside my nose. And it had long been agreed that I should have this done over the school Christmas holidays.

The Royal no longer exists as a hospital, but it was one of those splendidly old-fashioned city-centre infirmaries founded in the mid-nineteenth century, slightly intimidating and complete with classical columns. No doubt ill-adapted to modern medicine, it closed in 1997. But the Royal certainly knew how to do Christmas in style. For some reason, I had been allocated to an adult ward, and there were few patients left over the holiday season and no children at all. This meant that I was spoiled something rotten and not only by my parents (who lived thirty miles away and could not be there the whole time). By Christmas Day itself, I was well enough to enjoy a special appearance from Santa Claus bearing many gifts, a bedside performance from carol singers and a jolly visit from the surgeon who had done the operation (a Mr Clarke, I recall) – who brought his own daughters along for a bedside party. Who could forget that, almost fifty years on?

But, in truth, it was not the Christmas jollities that meant the most, or that I remember most vividly. It was the tiny little kindnesses done to a child in hospital for a few weeks and often on her own. My ward – Harper Millar – was presided over by the impressive Sister Gaye, whose deep-blue uniform dress, gleaming white apron and perched white, frilly cap I can still picture. And it was she who thought that I might like to take my favourite toy to keep me company on my way down to the operating theatre. This was a cheap stuffed version of Walt Disney's Goofy, and even now it is in my house somewhere. On the bottom of one of his flat feet, you can still just make out the words 'Mary Beard Harper Millar', which Sister Gaye insisted that we write, just to make sure that Goofy didn't get lost.

Every time Goofy emerges and I see those words, I remember what I owe to the NHS and to those who cared for me (whose names I'll never forget). Oh . . . and the nose? Well, the complicated engineering work was successful and I emerged as good as new, though – I admit – with a slightly blunted sense of smell!

RICKY GERVAIS

I was born at the beginning of the 1960s in Battle Hospital, Reading. I lived in a council house until I left home for university. I was the fourth child of an immigrant labourer. Men worked hard. Women worked miracles.

My mum was a homemaker and she could do anything except give me money. Luckily all the best things were free. Friends, family, nature, learning and healthcare. All of these allowed me to make the most of my life. That's why I gladly pay my taxes now that I've got a bob or two.

And that's why I love the NHS.

MAXINE PEAKE

Feeling Awful in Oxford

It was 2003 and I was on tour with a production of *Serjeant Musgrave's Dance*. We were nearing the final weeks of the run when we arrived at the Oxford Playhouse. We were a very close company and on the second day of our week's residency in Oxford, we decided to spend the day partaking in the activity of Laser Quest. (I know, I'm showing my age.) Basically, it's like paintballing's poor relation, although preferably less painful. At the end of the session I started to feel a little unwell. I put it down to exerting myself (a rare occurrence even now).

When I arrived at the theatre for that evening's performance, I started to feel progressively worse. The play is set in the Victorian period, so my costume involved a corset. I asked the stage manager if I could ditch it that evening as at this point I had already been sick.

He kindly agreed. Then, out of nowhere, I started to projectile vomit. In a break from my retching, I was questioned on my previous evening's alcohol intake, which had surprisingly been nil (I had a little crush on a fellow cast member who was teetotal, so I was trying desperately to impress him).

Anyway, I was finding it difficult to stand and the pain in my stomach was now excruciating. The company finally decided to call me an ambulance. I headed off to the John Radcliffe Hospital, Oxford. When I arrived, I was whisked into a cubicle where a very brusque doctor asked if the pain could be an STD? I looked at him a little taken aback and responded, 'I hope not.' At which he rolled his eyes and left.

I continued to be physically sick until the curtains flew back and a tall young man with a beaming smile entered.

'Hello, Miss Peake. How are you?'

'Not great, unfortunately,' I replied.

'Well now, if you would lie down on the bed, I'm going to examine you.'

Gratefully I lay down.

'Could you just turn on your side and lift your gown? I'm going to have to put my finger up your bottom.'

'OK,' I said. (Like you do.)

While the doctor was carrying out the examination he added, 'Oh, before I forget, I have Jane with me, she's a student doctor shadowing me this week, you wouldn't mind if she examined you too?'

'Er, not at all ...' I was brought up to always be accommodating where possible.

So Jane stuck her finger up my bottom too. The doctor

and Jane both thanked me and left.

Just as the curtain was settling, the tall young doctor popped his head back through, smiled widely again and said, 'By the way, I meant to say, I loved you in *Dinnerladies*!'

I was relieved to eventually discover I was suffering from appendicitis.

NICK HORNBY

Dear NHS,

When I was in my early twenties, I ruptured my cruciate ligament playing football and I had to spend a couple of nights in one of your hospitals. I was worried about not being allowed to smoke for that long but this was the early 1980s and when I came round from the operation, more or less the first thing I saw was that there was an ashtray by my bed and that the other patients, all young men with football or motorbike injuries, were puffing away. You were probably a bit behind the curve at that point, but your thoughtfulness was certainly appreciated. I was looked after by a very pretty young nurse and, after I was discharged, I was cheeky enough to send her a postcard, care of the ward, asking her out for a drink. (Reader, I … Well, we went out a few times.) Probably every single one of us would like our relationship with you to be something like that: a minor injury, a couple of fags, discharge, a few dates with one of your nurses.

But of course, it doesn't ever stop there. Life gets more complicated, more serious, scarier, and our connections to you become deeper and much more dependent. Ten years after my cruciate ligament injury, I became a father. My son Danny is severely autistic, it later turned out, but the autism came with a whole dizzying range of medical problems, many of them only obscurely linked to his condition. He was born in the Homerton Hospital in Hackney, but three months later he had an operation to repair a bilateral hernia at the Queen Elizabeth in Tower Hamlets. The granuloma on the back of his head was removed at Great Ormond Street when he was nine months old, and he was back there a few months later for an MRI scan for a pronounced left-sided weakness.

And then came the gut years, which have never gone away. He was a frequent visitor to the Royal Free – gut problems being a specialism of theirs – between the ages of seven and eleven, and he had surgery to insert an ACE tube when he was seventeen. In 2012, there was more surgery, to create an ileostomy, at the Royal London in Whitechapel. He was back there the following year for two lots of emergency surgery after complications. (The word 'complication' is rendered somewhat redundant by my son's medical needs and problems. He is a permanent complication.) Since then, he's been to A&E at the Whittington and admitted three times to UCLH. That's seven London hospitals for one baby, child, young man, now grown man.

I am lucky enough to have money to spend on private treatment for my children and perhaps some people may wonder why I haven't spent it. Why don't we get him off the waiting list, leaving you, dear NHS, more time and resources to treat those who can't afford it? Or – a question that might come from the particularly ignorant (or particularly fortunate) – why don't we want the best for him?

This last question is easy to answer: the NHS is the best for him. This is not ideological or emotional commitment speaking; it's a fact. My son's problems and conditions have been acute, yes, but also taxing, puzzling and sometimes right at the edge of the range and reach of current medical knowledge. In Britain, at least, the only people likely to know are the staff at the big teaching hospitals. We couldn't go anywhere else, because nowhere else would have been good enough. On occasions, there have been what seemed like entire roomfuls of brains thinking and talking about Danny. And we can forget this.

We all love you, NHS; we all think you are a great thing, that you have been underfunded, that, as I speak, you are being battered by wave after wave of desperately sick people, that when this is over we will want to thank you with more than applause. But we should also remember that the best and the brightest hearts and minds we have are all in your employ, and that is as important a reason to protect you as our sentimental feelings for you.

I shouldn't give the impression that my son's life has

been blighted by illness. He is happy a good deal of the time and he now lives in a wonderful independent living facility, with carers he clearly loves. We take him – took him, will take him again – to the pub a couple of times a week and he goes out with one of the carers to do his shopping every day. Who funds these people who are looking after him now, brilliantly, while his parents can only follow his progress from a distance? You do, NHS. He would not be there, or maybe anywhere, without you.

MARK HADDON

Every single member of my family owes their life to the NHS. I take this for granted, and only at moments like this when I pause to consider what might have been do I feel a heady mix of vertigo and gratitude. Most of those stories are not mine to tell, so let me tell just two of them.

In 2019, I had an unexpected and, in my opinion, somewhat undeserved triple heart bypass at the John Radcliffe Hospital in Oxford. A long vein from my left leg is now doing a slightly different job inside my chest and I can still feel the bumps of the six wires used to hold my sternum together after it was sawn in half from top to bottom, but I can run happily for hours up and down hills of all sizes. I feel as if I have been given a large second helping of life.

The fact that the human heart can be replumbed is astonishing. Even more astonishing is the way human beings have organised themselves to make such miracles possible. I owe my life not just to my GP who sent me for a CT scan and to the surgeon and his team, but to a long human chain of people who looked after me before, during and after the operation – nurses, porters, cleaners, doctors, technicians, care assistants …

My wife, Sos, would doubtless say the same about her own experience of the NHS when, in the summer of 2003, she had a near-fatal bike accident in the Black Mountains, though her list would include some glamorous paramedics in red jumpsuits and a helicopter pilot. She was six months pregnant when she hit the bonnet of a 4x4 on a tiny country road and I really did think I was kneeling in the road watching the person I love most in the world dying in front of me.

She'd broken her pelvis in two places, fractured her cheek in three and one side of her face was composed entirely of bruises, swelling and road rash. The baby was alive but not moving. She spent only seven days in Bristol Royal infirmary because she is a very tough woman indeed, and she walked without crutches for the first time after having an epidural in the birthing suite.

Our son is now sixteen and appears completely unaffected by his in-utero adventure and the large servings of morphine afterwards.

I've told both these stories a number of times over the years and I get a little choked up every time. I get choked up talking about the overworked and underpaid staff who nearly always provide an amazing service with dedication, kindness and patience. I get choked up describing how the NHS is one of the most principled, humane, generous and efficient institutions in the world, and one of the few remaining reasons I feel proud to be British, despite the drip, drip, drip of privatisation by politicians and businesspeople wanting to push it gradually closer to the hellscape of an American medical system where people go bankrupt because they have cancer.

But there's a coda I always add to both stories, and it is this which never fails to bring tears to my eyes.

I love the NHS most of all because we pay for it with our taxes, and because the care we receive is the same whether we've paid a million pounds or nothing. We don't talk enough about this. We all need healthcare; it's expensive and we have to pay for it one way or another. Serendipitously, the fairest way of doing this is also the most efficient. If we want to support the NHS we need to celebrate tax, to think of it not as money the government steals from us but as our contribution to a safe, just and healthy society. We need to think of tax evasion and tax avoidance as stealing from hospitals and schools, and we need those who are well off to pay more. And I include myself in that category.

The true worth of the NHS is not that they saved the lives of every member of my family. It is that they make the same effort for all families, even if those families are destitute. The true worth of the NHS is that those of us who are lucky enough to pay tax can go to sleep at night knowing that we have helped make that radical kindness possible.

KATE TEMPEST

We're held at birth, and held at death
And all through life. *Don't hold your breath.*
Just count down backwards – five to one.
Despite ourselves, the pain has gone.
You face the monsters we all dread.
We tightrope walk across the depths
Because you go one step beyond
To hold up that vast safety net.

Double-shift, now watching dawn
Flood the car park, sunlight warm
Against your lids. All in your stride;
The blood, the mess, the tears cried.
The people lost. The babies born.
When we are at our lowest ebb
Scared and weak, our spirits fried
You dress our wounds and change our beds.

You meet the worst and best in us;
The perseverance, will to trust.
The brave, good-humoured, got-to-laugh
The angry, drunk, the-can't-be-arsed.
The curious, the wild, the crushed.
The out-of-patience, worn-out, stressed.
Some take for granted, some take charge.
Most are grateful. And impressed.

Now, hot drink and pounding head
Must get to sleep, must get some rest
All starts again in seven hours.
Digging deep for hidden powers.
By the gates, the air is fresh;
The street is sweet with sudden Spring
I see you stop to smell the flowers
Deep breath out, then head back in.

It's good to stand applauding yes
It warms us all, but begs the question –
Is there not a case to make –
To give the NHS more headroom?
Hailed as heroes in the press
As if you're on some driven quest
Instead of people working jobs,
You do the things you do because
You have to do the things you do.
Funds are cut across the board
And you are left to hold the fort.

So here's to you,
The nurses, cleaners,
Admin staff, healthcare assistants,
Paramedics, drivers, porters,
Volunteers and heavy lifters –
May you find the strength you need
To stay the rounds and go the distance.

CHARLIE BROOKER

I was born in an NHS hospital. At least, that's what they tell me. Call me forgetful but I have no memory of any of it. I could've been born in the log flume queue at Chessington World of Adventures for all I know. Or in a lowly cattle shed, like Jesus. Which would explain why I take after him in so many ways.

Both my kids were born in NHS hospitals too. The first one tried to show up early. Several weeks before the due date, my wife Konnie shook me awake at 6 a.m. to say she was a bit worried her waters might have broken. She couldn't be sure. She'd just gone a bit ... leaky.

My brain ran around screaming, but my face remained calm. The very picture of denial, you might say. We decided to wait until 8 a.m., when our local surgery opened. It was just around the corner. We were halfway there when the trickle became a flood. The GP sent us straight to St Thomas's Hospital in a cab. We expected the baby to pop out at any time. Instead, Konnie was kept in for about a week, for observation, before being sent home with several sheets of disposable strip thermometers and instructions to return immediately if her temperature rose or the baby decided to appear. If neither of those things happened, they'd induce the birth at thirty-seven weeks.

The temperature stayed put and so did he. Who knew a baby could remain indoors so long after the waters had broken? No wonder he coped with the Great 2020 Lockdown so effortlessly (at least at the time of writing – if the Nintendo Switch breaks, it's going to be like sharing a cell with a miniature werewolf).

Come the thirty-seventh week, we were back at St Thomas's for the season finale. We got there bright and early at 8 a.m., but it turned out they were having a rush on and we'd have to wait. After spending almost the entire day in a room full of women clutching their swollen bellies and groaning, like the aftermath of some insane Christmas feast, they led us to a room and gave my wife some drugs to induce the baby. We were to stay in the room overnight. Like a dumbo, I hadn't envisaged staying this long and was ill-prepared. They found me a sort of yoga mat and I slept on the floor, like a divorcee camping out in the garage.

By the next morning there was still no sign of the baby. Hours stretched by. They prepared Konnie for an epidural and I played *Angry Birds Space* on my phone. It's amazing how many details of this kind of life-changing incident you forget; I don't know the names of any of the medical staff who helped us, yet I can clearly remember that *Angry Birds*

Space – which is like *Angry Birds*, but in SPACE – had literally just been released that morning, because it took forever to download from the App Store using the public Wi-Fi.

Suddenly, the machine that was monitoring our unborn baby's heart rate made a noise I didn't like and moments later the bed was surrounded by the entire cast of *Holby City*. I was handed a set of scrubs and told to go and change into them because they were going to perform an emergency Caesarean any moment.

I nodded calmly, kissed Konnie on the forehead, strode into a disabled toilet to get changed, shut the door and instantly found myself clinging to the handrail like I was in a rocking carriage – because my legs had become noodles. I slowed my breathing, tried to calm my internal organs, wept efficiently for five cathartic seconds, then began wrestling with the scrubs, which turned out to be about five sizes too big. Also, I had no idea how I was supposed to put them on.

Moments later, I emerged, looking like a schoolboy in an ill-fitting Halloween surgeon costume. My heart rate was through the roof but, in the moments I'd been absent, the baby's heart rate had thankfully stabilised. Another doctor had arrived and the new consensus was that the emergency was averted – or at least postponed. They wanted to wait a few more hours in the hope that he'd eventually come out on his own. They were just like the FBI surrounding David Koresh's compound at Waco, except none of them chewed gum or said, 'Shit's about to get real.' Not within earshot, anyway.

Konnie tried to snooze. I tried to take my mind off things

by using my phone as an intermittent escape hatch – but let me tell you, the difficulty of *Angry Birds Space* really intensifies when your fingers feel like electrified bananas because your adrenalin is still at wartime levels.

And then, forever later, but also somehow instantly, Konnie was being wheeled into the operating theatre while I tagged along. I was jittery; she was out of it on whatever new drugs they'd pumped her full of. She's teetotal, so part of me enjoyed watching her slur her words and loll her face around like a soap-opera matriarch with a drink problem for once. In fact, I wish I'd filmed it.

I stood beside her and held her hand while they raised a sheet at around tit level so we couldn't see them cutting her open behind it. I remember trying to stroke her hair in a manner that struck me even at the time as possibly a little patronising. Sure enough, even through her narcotic haze she told me to stop because it was doing her head in.

What happened next was like a magic trick. They rummaged around behind the sheet, delving into her insides while somebody somewhere recited medical jargon, like some occult incantation. There was a surprising amount of pulling, pushing and wrenching – and then suddenly someone was holding up a baby, like a magician pulling a rabbit from a hat. A skinned, screaming rabbit. Which we had to take care of for the rest of our lives. It should have been nightmarish. But it wasn't. It was great.

For months, I'd been secretly worried I might feel nothing when the baby was born. After all, I'd previously never really 'got' children. They'd always struck me as an inherently unknowable species – like cows or parrots. Cute, I suppose, but who really gives a fuck, right? And then, when it came to it, when it happened . . . the moment they put this warm, vulnerable infant in my hands, I felt my brain rewire itself with a new sense of purpose. Love. Love and protection.

I was instant pudding.

'He's beautiful,' I said to Konnie.

She nodded, woozily, her eyelids half shut.

'Do you want to hold him?' I said.

Instead, she asked for something to be sick in.

A few hours later, we were on a different floor of the hospital, our son wrapped in a towel with the word 'SUNLIGHT' printed on it, lying in a sort of Perspex crib-trolley beside Konnie's bed. She slept off whatever they'd given her, while I took 500 photographs of the baby.

St Thomas's Hospital stands beside Westminster Bridge. Our son's crib was positioned directly beneath a window facing the Thames. Night had fallen. Big Ben glowed across the river. The House of Commons stood illuminated, golden. The building where, sixty-four years earlier, the National Health Service Act had been passed.

I took a few steps back and snapped a picture of our firstborn sleeping there, with Parliament visible behind him through the window.

I'm no good with cameras. The picture I took is poorly framed and quite blurry, and by far my favourite photograph of all time.

ANNE FINE

Times change, as do social expectations and NHS guidelines. So though I've had my share of broken arms and weird infections and family members who are still with us only because of medical and nursing skills, it's something else, from over forty years ago, for which I'm most grateful.

I went to see Dr R with my grizzling toddler and chatted while he examined her. 'I'm going to apply to medical school,' I told him. 'My degree is in politics, but that's not very useful, is it? I'm going to train for something else.'

He nodded at my daughter. 'Not thinking of having another, then?'

'Not really, no.'

He said, 'You realise that babies are like puppies? Better off with company. Take my advice. Pack in your plans and have another baby instead.'

No doctor would give you that advice these days. And if they did, they'd either find themselves on the floor, or struck off the register. But those were different times. I went home and got pregnant. I've never told anyone about this before now, but we all love the NHS for different reasons.

And my very, very precious second daughter is mine.

SHAPPI KHORSANDI

'Just to warn you, Shap,' my brother stepped out to tell me before I went into the ward, 'there's a bucket of, well, blood by Dad's bed.' I strode in to give my father strength and support after his triple heart bypass operation. I saw the tubes coming off him and into the bucket. Luckily, his bed cushioned my fall as I fainted.

The NHS has mended my father's heart several times, and attended to quite a few of my fainting spells. I am not usually the first to be despatched to the bedside of ailing parents. When my mum had three epileptic fits in one day, I happened to be the only one in our immediate family who was in London to take care of her. After the first fit, I took her to my house where she had another in my bathroom. She fell and cracked her head on the floor because I know nothing about epilepsy and had left her in there alone. The hospital was rammed; nurses were rushing about like Roadrunner. It was impossible to stop one even for a moment to ask them about my mum. When the doctor came to see her, he asked questions to check how lucid she was.

'When did the Second World War end?' he asked.

'Yes,' my mother replied.

I said, 'To be honest, Doctor, I'm not sure my mother would know the answer to that on a good day.'

He tried another. 'Who do you live with?'

'Alone,' my mother answered immediately. I reassured myself, 'That doesn't mean anything, it's easy to forget she has lived with my dad for forty years, he's quite small after all.'

My mother turned her head. Haunted terror spread over her face as she looked at something invisible, petrified. Then her eyes rolled back, she let out a wail and began to fit. Those nurses who didn't have a second to stop and talk to me were, in a blink, at her bedside, holding her as her body shook violently. I stood back, frantic, and could only watch as this little army took over. They'd heard the sound that came out of my mum and knew they were needed and absolutely knew what do to help her while I stood uselessly gawping as she jolted and convulsed. The fit seemed to go on and on. The nurses looked after the woman I cannot imagine life without.

Once she was home and recovering, my mum had no recollection of the doctor's questions, so I told her. In my thank-you card to the staff, she made me write, 'My mother would also like me to assure you that of course she knows when the Second World War ended, and she may as well live on her own for all the work he does about the house.'

My two children had a soft landing into the world, thanks to the NHS. My first, a son, was a natural birth. The pain was a shock. I couldn't understand how I could be in that much pain but still be alive. I screamed at the midwife, Cynthia, 'GET ME AN EPIDURAL OR A GUN.' Happily, they were out of guns and Ahmed, a good-humoured anaesthetist, arrived, injected some manners into me and I apologised to Cynthia.

A few years later, my daughter came by Caesarean section. Like my son, she was greeted into the world by the United Nations of the NHS, with English, Welsh, Irish, Pakistani, Nigerian, Jamaican and Filipino staff all on hand to care for us. I had her by myself; obviously there was a man involved at some point, I am not a worm, but he did not turn up to see her. Very rude of him considering it happened to be MY birthday too. The midwives and the doctors were extra-attentive, knowing my personal soap opera, holding my hand as I sobbed, making a nest of feathers for me and my baby girl and letting me watch *Pretty Woman*.

No matter how baby-proof you make your home, they will miraculously find a piece of barbed wire or a chainsaw from somewhere and play with it. There were trips with both children to mend fingers and bandage heads, and once, when my daughter was four, an operation to remove a one-pound coin from her oesophagus. I have never had to be as strong as I was in that moment when I held my child's hand and watched as she was put under general anaesthetic. The anaesthetist saw how frightened I was. He took me aside and looked me in the eyes and assured me my baby would be fine. Despite the hundreds, maybe thousands of times he'd done this, he didn't dismiss my fears, he understood that my entire universe was in his hands. He explained the 'sleepy milk' to my daughter and put her at ease. To see her tiny body and expressive little face succumb to a drug and become unconscious sent my heart out to every parent who has had to see the same, again and again, for those for whom hospital stays are part of life.

As my sleeping child was taken away to theatre, I thought of these doctor and nurses who look after desperately poorly children, talking to parents who are frantic with worry or wild with grief. Now here they were, soothing my worries with the same compassion and professionalism over a routine procedure, taking the most exquisite care of my human money box until she bounced happily out of the doors just a few hours later, covered in stickers, high-fiving doctors, excited about the ice cream she'd been promised. Thank you, NHS.

STANLEY TUCCI

A few years after my first wife passed away, I married a Brit and ended up in London with my three children. Needless to say, there were many adjustments to be made: the all too well-known driving on the left side of the road issue in a car that had a 'bonnet' and a 'boot' instead of a 'hood' and a 'trunk'. This was only the beginning of the language disparity. I didn't know what a 'rasher of bacon' or a 'banger' was, nor did I know what it meant when someone was being 'shirty'. I was clueless as to the definition of a 'skip' or a 'tip', or that when I was wearing a 'sweater and sneakers' I was actually wearing a 'jumper and trainers'. Having lived in NYC for many years, I most certainly wasn't used to encountering cab drivers who were polite and actually knew where they were going. And, to be completely honest, I didn't know what a 'willy' was. Don't worry, I know now. My wife explained it to me. And now we have two kids of our own.

But the aspect of the UK that took me the longest

amount of time to become accustomed to – and even *comprehend* – was the fact that I or any family member could go to our local surgery and be treated without anyone asking for payment, either upon entering or exiting, and that no bill for services rendered would ever arrive in the post. When one of my older children cut her heel rather severely while bicycling, she was treated immediately with great care and attention, continuously until she was healed. When our youngest son twisted his arm while we were visiting friends out of town, the NHS doctors at the small hospital we took him to were not only professional but extraordinarily gentle with him and reassuring to the two panicked parents, namely me and my wife. (For the record, it was her fault.) Again, I was amazed that no one asked for an insurance card, a credit card or any form of remuneration. I still expected a bill to arrive in the post weeks later. Needless to say, one never did. That diligence, professionalism and kindness are what I have experienced over and over again when any of us have been treated by the NHS.

I am not saying that these qualities don't exist in healthcare professionals in other countries. They do. I have had extraordinary doctors and nurses in America over the years. But my inability to reconcile how this system works is exactly because I am from America, a country that in 2008 had over 44 million people uninsured. Thanks to Obama's Affordable Care Act, that number has dropped to about 28 million but is steadily increasing again as conservatives insidiously roll back the laws that made the positive difference in the first place because they see them as socialistic.

I explain all this so that people might understand how, no matter how inefficient and frustrating dealing with the NHS might be from time to time, it is one of the most extraordinary healthcare systems in the world. Can it be improved? Yes. But from what I can discern, any troubles it may have stem from a lack of proper and consistent funding. Spending my time and being a taxpayer in both the US and the UK, I've come to understand how precious and vital the NHS is. Not only for the physical and psychological health of every UK citizen, but for the economic health of the nation as a whole. The healthier a nation is, the more productive it is, the safer it is and the happier it is.

The people on the ground in the NHS who are working day in and day out during a crisis such as we are facing now are risking their lives to save countless others. Any politician who thinks of privatising the NHS should have their head and heart examined, and they should pay out of their own pocket to do so. Sadly, it may well take this crisis to make them understand that the doctors, nurses and staff of the NHS are national treasures as much as the institution itself. May it have a long and healthy life.

DERMOT O'LEARY

My earliest experience of the NHS and the care it provides was as a small, slightly confused boy of nine.

My dad, who had cancer of the tongue, was in hospital in Colchester for what, at the time, seemed like days; it could have been weeks. All I remember is not being able to visit him but seeing the room on the first floor from the street below and waving to an empty window as we walked by. Then, him tired and unshaven at home on the sofa, groggy but happy. My dad was never unshaven; he had, and still has, the energy of a twenty-year-old and so to have a nap halfway through the day then was unthinkable. Hence the slight confusion on my part. But that, thank God, and the NHS, was that. He recovered and life went on.

Or so we thought. As so many who have had the big C find, it tends to return with interest.

Five years later, the cancer came back and this time an operation was needed. After a gloomy, life-altering prognosis from a local doctor, Dad demanded a second opinion and – indicative of the people my parents are – they went quietly and diligently about finding someone who could, just might, offer an alternative.

My parents have never shielded my sister and I from anything much. Being the youngest, I might have been cushioned from the worst of news but they've always treated us with the ability to rationalise and think through problems. That said (and now being an adolescent who could process slightly more . . . slightly), as much as I knew was that it was serious. My main memory from the time as a bleak winter set in were windswept and rainy walks on beautiful but desolate beaches in Essex, as they walked ahead, having the 'What if' chat.

We were a Catholic family – still are – but faith has always been used as a compass, a guide to plot your path, rather than a vessel of easy certainty and swift judgement. So when that prognosis came, yes, there were prayers involved and faith was leant on, but to equate one's level of faith and devotion to whether someone deserves to live or die has always seemed fruitless and unhelpful, and a bit mean. So, prayers, yes, but also a second opinion.

The sought-after second opinion was found and, before we knew it, my father was in that great Victorian outpost of hope in the fight against cancer, The Royal Marsden. We'd be in London over Christmas, staying with our ever-supportive Irish family.

The treatment was a hemiglossectomy with repair, a very new treatment. In essence, removal of half the tongue and a rebuild, utilising the wrist pulse to provide a blood supply. Drastic, yes, but also brilliantly innovative and meaning, if successful, that my dad wouldn't lose the power of speech. A date was set and the procedure went ahead.

The tubes and the beeping, that's what hits you first – well, second, after the clean smell, that smell that is at once both foreign and familiar. But it was the tubes going in and out of my dad – my Everest, my redwood, my lighthouse – I hadn't expected. I don't know what I had expected; maybe a fourteen-year-old boy (or this fourteen-year-old boy) didn't think enough about what to expect. But it hit me, punch-on-the-nose hard.

Dazed and with my legs giving way I'm led out onto the street by my uncle Frank, my quiet, wise old uncle Frank, who gives me enough time to compose myself, whilst (with obligatory Superking screwed into his mouth and lit ... what does the brilliant Editors' song say? 'The saddest thing that I'd ever seen / Were smokers outside the hospital doors') he tells me my dad will be OK. Early days, certainly, but the operation was a success.

And it was. An operation which, after speech therapy and excellent care by the NHS team at The Royal Marsden, led to a full recovery – thanks to a second opinion and the forethought, skill and compassion of NHS doctors willing to take a shot and not to settle, instead to look forward and to innovate, all with the patient's best interest at heart.

My dad ended up becoming a very unlikely poster boy

for that kind of surgery, the procedure being pioneering at the time, and he's still here now and, touch wood, cancer-free. So, thank you NHS, for the gift of my father.

PETER KAY

Car Share

John and Kayleigh are driving home from work
Gina G – (Ooh Aah) Just A Little Bit is playing on Forever FM

JOHN: (*Laughing*)

KAYLEIGH: What are you laughing at?

JOHN: I'll tell you an embarrassing story about this song.

KAYLEIGH: What?

JOHN: I got a call once from one of my dad's consultants at the hospital and for some reason I happened to be playing this song on my phone.

KAYLEIGH: Are you a fan? I love this song.

JOHN: Of Gina G? Give me some credit. I think I was setting up Spotify on my phone or something and this song randomly came on, then the phone rang and I couldn't turn it off. So there I was having this really serious chat about my dad's palliative care plan with 'Ooh Aah Just a Little Bit' playing underneath. The consultant said, 'Who's that singing?' and I just had to pretend I couldn't hear anything.

KAYLEIGH: How embarrassing.

JOHN: I tell you another one worse than that. I went for a brain scan once and I was . . .

KAYLEIGH: Oh my god, why?

JOHN: It was years ago, I kept getting these really bad headaches. My GP said it was my sinuses, but being a proper hypochondriac I thought bollocks, what does he know? So I booked in for a brain scan.

KAYLEIGH: Where, at the hospital?

JOHN: No, B&Q car park.

KAYLEIGH: There's no need to be sarcastic, John.

JOHN: I'm not being sarcastic! That's where it was. They have these big NHS trucks that pull up and they do scans in the back of them.

KAYLEIGH: I've never seen those.

JOHN: They're there all the time, parked up behind the Shell garage. They do all sorts. One of our Paul's mates had a vasectomy and he'd only nipped to B&Q for a bag of nails.

KAYLEIGH: Ouch.

JOHN: Anyway, I went in and met this nurse, and she asked me if I wanted to listen to any music while I was having the scan and she showed me a load of CDs . . .

KAYLEIGH: What did they have?

JOHN: Frank Sinatra, Elvis, Best of Steps . . .

KAYLEIGH: I'd have gone for that.

JOHN: Of course you would. I went for the best of Simon & Garfunkel.

KAYLEIGH: Urgh.

JOHN: I love them and they would have been the perfect choice if the first song hadn't been 'The Sound of Silence', with the opening line, 'Hello darkness my old friend, I've come to talk to you again.'

KAYLEIGH: Nooo!

JOHN: Yeah exactly . . . and there's me laying in this MRI machine having a brain scan. It got worse, in the third verse the lyrics are 'silence like a cancer grows'. I was in tears. I was ashen when I came out. I told the nurse. I said I thought it was going to be happy, Mrs Robinson and all that. She said a lot of people pick Frank Sinatra because they think it's going to be all 'Fly Me to the Moon' and 'New York, New York', but the first song is 'My Way' with the opening line, 'And now, the end is near and so I face the final curtain.' You'd be gutted.

KAYLEIGH: And so did they find anything?

JOHN: What, like a brain?

KAYLEIGH: No, I mean was anything wrong?

JOHN: No, nothing. The GP was right, it was my sinuses after all. A course of antibiotics and I was sorted in a week. That'll teach me.

JOJO MOYES

Like most people who grew up in this country, my life and my history are inextricably entwined with the NHS. I was born in 1969, ten weeks too early – a figure that should have heralded a death sentence. The pictures of me from that time – a translucent baby bird – are hard to look at. And yet somehow, with an incubator and the Herculean efforts of doctors and nurses, I thrived.

My childhood was a series of corrections and tweaks overseen by its myriad services: ophthalmologists (thank you for my glasses), dentists (thank you for my straight teeth), urologists (fixed that recurrent kidney infection), radiologists (yet another broken bone, set and repaired), and my adult life has been punctuated by its intermittent and less obvious favours: antibiotics, smear tests, anti-depressants and – gift of all gifts – the safe delivery of three beloved children. For whom the same cycle has been repeated.

When we had our third child, our good luck in being

born here was brought into sharp focus. The fact that he was profoundly deaf was picked up at a specialist screening when he was just two weeks old and then confirmed after more comprehensive testing, one doused with tears, at eight weeks. That crucial early diagnosis (many deaf babies and toddlers become so good at imitating that parents often refuse to believe they cannot hear) meant that he was fitted for hearing aids at six months and that at fifteen months he was offered a cochlear implant – a coin-sized device that would be surgically implanted into his head and replicate the connection between ear and brain.

That three-hour operation, its attendant appointments and equipment, which took place fourteen years ago this month, would have cost £46,000 – a staggering amount and one which we would have had to remortgage to find. But because of the NHS, our son was given his hearing for free.

That it has worked beyond anybody's expectations – allowing him to attend a mainstream school, enjoy music, talk over his siblings at a dinner table and even mimic people's accents – is something I am struck by daily. If he chooses to not wear it and be part of the deaf community, that will be fine too. I'll always be grateful beyond measure to Patrick Axon of Addenbrooke's Hospital who changed all our lives with his surgical skill (and beautiful, tiny stitching) for giving him that choice.

But oddly, those achievements are not what I remember most when I think back to that time. I remember the kindness of a nurse who supported me when my baby was placed on the operating table and fought the anaesthetic. My legs buckled as I walked out, and she caught me and brought me tea and toast ('the mothers never have breakfast'). It was the greatest tea and toast I have ever eaten. I remember the smile of the senior nurse in Audiology whose calm manner helped allay my worst fears. I think of the engineers who helped 'switch him on' six weeks later and shared in our joy when our son heard his first sounds, his eyes wide with wonder.

We might have taken the NHS for granted once, but not now. Not in the depths of a pandemic where the heroism of frontline staff – often operating without the right protections – humbles every day. But also when we look across the Atlantic and see those who do not have the same healthcare. Those bankrupted by cancer diagnoses. Those who have spent five or six-figure sums just to safely birth a child. Those like the woman I met recently on a plane who had spent a million dollars in excess of her top-level insurance policy just to care for a severely diabetic and disabled husband.

But I believe the NHS serves a greater purpose in the British psyche than just caring for our health. It reminds us that there is such a thing as the common good, an idea of true service. And this is what I take from mine and my family's experiences with the NHS – that sense of service, something way beyond the desire of making money or progressing up a career ladder – and that sense of goodness.

Thank you, thank you to everyone who works for it.

DAME JULIE WALTERS

Dear NHS,

You may not remember, but we were an item some years back, 1968/9 to be precise. You always loved a uniform and I gladly went along with your demands, donning my starched white cap and apron to please you on countless occasions. Alas, our love affair was not meant to last, and you threw me out following an incident with an elderly man and a bedpan, whereupon I was hauled up in front of the matron after his relatives threatened to sue.

Anyway, I just wanted to say, I never stopped loving you, despite the rejection, and now you are my total hero! Without wishing to sound like a stalker, my love burns ever brighter and will never, ever end! And yes, it was me hanging about in the bushes outside A&E last night and shall be again and again until you see the error of your ways.

Eternally yours,

Nurse Julie Walters ♥ ♥ ♥

MATTHEW SYED

I had never before sat at a person's bedside as they passed away but a few years ago I was at the hospital beside Philip, my beloved granddad, as his breathing became quieter. His deterioration had been quite steep, perhaps understandable given that he was nearing a century in age.

Beryl, his elder daughter, was hurrying from Heathrow airport where she had landed from her home near Washington DC. His younger daughter, Dilys, my mum, with whom he had lived for the last two decades of his life, was rushing back from a rare holiday.

I couldn't quite believe it was the end. He had been a mentor, a confidant, the dearest of friends. I thought of the many visits to the home he shared with Nana, who had passed a few years previously, up in Prestatyn – our walks along the seafront, our long chats in front of the fire, the way he would offer advice. Tears came and went as I looked at his face, peaceful, resigned.

I was conscious of something else too. The nurses. It

wasn't the fact that they checked his vital signs, monitored his breathing and made sure he was comfortable, even while semi-conscious. It wasn't the diligence that they showed in everything they did. Rather, it was the compassion that went beyond professionalism, a concern for this unique human being, my granddad, despite the many other people on a busy ward, struggling in their different ways.

A nurse brought me a glass of water. She had brown hair and deep brown eyes, and I agonise to this day that I don't remember her name. She explained what was happening, explained that he was near the end, and offered wise words that might help me to cope with the finality of it all. As I listened, I noticed that his breathing was subsiding once again.

Almost as she finished speaking, he took a deep breath. I shot a look at the nurse, worried that he might be experiencing pain. She explained, patiently and kindly, that this was normal in his condition. Just then, Granddad seemed to stop breathing altogether. I leaned forward and put my ear to his mouth, caressing his head. I stood back upright, alongside the nurse, my heart pounding. 'He's gone,' she said gently.

Time is such a strange and elastic thing. In my memory, I stood there for many minutes taking in her words but it can only have been a few seconds. I shall never forget the nurse leaning down and kissing Granddad on the forehead. 'I am not supposed to do that, but he was such a wonderful man,' she said. 'You get to know each of your patients personally.' Despite my grief, I was astounded by her humanity. 'Thank you,' I murmured.

Beryl came into the room perhaps five minutes after her father had died. She had rushed from Heathrow and sprinted up from the hospital car park to the ward. We embraced. She had worked as a nurse her entire professional life, towards the end of her career at the Royal Hospital in Chelsea where she had cared for the old and frail. But she too was profoundly moved by the compassion of the nurse, who returned multiple times to see that we were OK, to offer her help, to take us by the hand.

'You have not lived today until you have done something for someone who can never repay you,' John Bunyan wrote. I will never be able to repay that nurse, except by writing these words and reminding her, if she should ever read them, that the grandson of Philip Heard, who died on 11 January 2013 at the Royal Berkshire Hospital in Reading, was helped immeasurably by you.

You were there for Granddad, there for me, there for Beryl and there for the thousands of other patients who have had the good fortune to cross paths with you; a tapestry of help, assistance and care that cannot be adequately measured but which should be more fully acknowledged.

Thank you.

SALI HUGHES

In 2016, I spent the best part of a month in a hospital, having been terrified of them my whole life. There was a childhood incident when a dermatologist scooped out a lump of my arse cheek and popped it in a screw-top tube. I got dizzy, fainted, and when I woke up, my mother was screaming needlessly and mortifyingly at staff. I barely went back for the next thirty years. Instead of visiting sick friends, I sent lovely treats, thinking they were as valuable. I gave birth to my babies on my living-room floor (for many reasons, but chiefly because hospitals made me extremely anxious), I performed minor home surgery wherever possible, avoided A&E unless it felt life or death and consequently managed to swerve even a single night in a mechanical bed. Hospitals were where you died or saw people dying and I wasn't going near them. Of course, as phobias go, this is something of a non-starter. Everyone ends up there in the end. And so it was, a few years ago, when my youngest son was unexpectedly taken very ill and admitted to hospital and, by extension, so was I. Time to woman up and be with my child.

It was to be a defining month in my life. The NHS doesn't just change the lives of patients, it also, I was to discover, changes those of their families. It causes a wholly unique sensation of dread and longing, in that you visit as a result of danger or jeopardy and would very much rather not be there at all. And yet everyone is there to help you, to reassure, heal and comfort. They're not anonymous uniformed beings and you're never just an NHS number. The service is made up of ordinary, yet extraordinary, humans who walk among us but operate on a much higher plane. The male nurse who always tells your kid how chuffed he is to see him, however many hours he's worked on the trot. The female nurse who kneels alongside you, cheerfully scrubbing poo off the floor while congratulating the whole family on the unblocked bowel. The junior doctor who stands his ground when no one above him agrees it's appendicitis; the anaesthetist who touches your hand as it leaves your unconscious child's leg and tells you to 'go and anaesthetise yourself in the pub'. The surgeon who decides to prescribe the super-duper antibiotics ring-fenced for cases of life and death – the drugs that saved you from the hell of more surgery. The porter who brings an extra blanket to your chair while you're sleepily reading your Jilly Cooper, one eye on the heart monitor. Even the permanently furious woman who takes the lunch orders while appearing to hate adults, children, food and life so much that it provides your whole family with their first laugh in days. These are dutiful, kind, clever and immensely caring people you grow to sort of love, despite hoping that, soon, you'll never have to see them again.

But on it goes, and you sit helplessly in awe. After almost three weeks, my son became the ward's longest-standing resident, but was nowhere near the child in most dire need. There were caterwauling babies on drips being walked up and down corridors by nurses pushing prams with one hand and IV drips with the other, softly shushing into ears the size of pasta shells. There was a child with no limbs, several toddlers restrained from toddling by a host of tubes and pipes, teenagers in waiting rooms who must have no sooner got their GCSE results than had to embark on chemotherapy. The noises – children coughing, alarms beeping – that once made you wake in a panic quickly became white noise, just another malfunction of tired equipment limping stoically through another year.

Hospital soon becomes life. Your new world is tiny and insular – because little else matters when your child is ill, but also because the phone signal is pants and hospital WiFi is wonky. Events which history will classify as incidental are, in any case, rendered trivial by circumstance. I remember hearing about the Brangelina split some five hours after the internet and having no thoughts beyond 'poor them' – and even that had already passed before the next doctor's round. I missed the futile challenge of Owen Smith for the

Labour leadership and – despite being a party member and sitting in an NHS hospital ravaged by Tory government cuts – I struggled to care. A sweaty, tube-tangled cuddle and the barely-there plot of *Blades of Glory* played out on a TV screen bought with public donations seemed a better use of my time. In hospital, your usual interests vanish. Anything but getting your family well seems like extra baggage on the run.

And then one day you're told you can go, that the hallowed drugs have worked miracles. You'll no longer sleep fitfully in your pull-down single bed, your child will be relieved of his nasal tube and three separate cannulas and released from his prison to go home to his Xbox and dog. And so you pack your suitcase of Get Well cards, books and free disposable thermometers, you say your thank yous and deliver your hugs and you just walk out – better, happier, healthier, no longer terrified nor even a penny worse off, despite the fact that, just a week ago, you'd have gladly handed over your house deeds to anyone who could stop the foreign, blood-curdling wail coming from your own child. They don't care who you are, what you earn. There are no bills; they're already settled for an average of two grand per person, per year – roughly the same as we splash annually on takeaways and around £4,500 *less* than the average insured American spends on their private healthcare (not the other 70-odd million who just go broke or bankrupt, avoid going to the doctors when ill or, in 45,000 cases, die needlessly). Your ingrained routine built around doctor's rounds and drug administration is simply adapted for the next person; your grey, wipe-clean room is given to another story, another family to save.

And as you're catching your breath, enjoying feeling real outside air on your face and thinking that to be British is the luckiest imaginable fate, the hospital staff have already gone back to doing their thing – taking broken children and adults, opening them up, putting them back together, shrinking the aggressor, dressing the wounds, cleaning the sores, stroking the faces and breaking the news, like any of it is normal. And all while wading through the treacle of brutal cuts, unfathomable reorganisation, debilitating admin and a whopping broken Brexit promise that is already considered dreary and whiney to bring up. I'll never fear NHS hospitals again. Only feel tearfully proud of their heroism.

KIT DE WAAL

Never Forget

Weekends were mother-free. She volunteered to work Friday, Saturday and Sunday nights, the most unpopular shift at the big old hospital that covered the poor, immigrant, working-class north of the city. She was an auxiliary nurse but we never really understood the difference between that and the real thing. She wore a beige uniform and a little upside-down watch. She wore man-like lace-up shoes and always smelt of disinfectant when she came home in the morning, ashen-faced sometimes at what she had seen.

'A little thing was born last night,' she would say and then whisper, 'hardly a baby at all.'

We winkled the details out of her – the missing limbs, the extra limbs, the strangeness. But it became too terrible when she described the mothers' pain, the wailing and endless agonies of labour, so we would wander away from her mid-sentence to the television or garden.

She was a little Irish woman built for rabid industriousness.

It suited her personality – hefting patients, straightening sheets, scrubbing, cleaning and making things right. She'd walk through the door at breakfast time and start the washing-up, telling us about her night's work and casually drop in phrases like 'bed pans', 'fifteen stitches', 'rectal tear' and 'bloody sanitary pads' and we, teenage girls, would gag and shudder.

With Christmas came a bumper crop of chocolates and biscuits, and there were year-round flowers and thank you cards from the mothers who made a special trip back to the ward to thank her specially, bringing the swaddled infant and grinning father. She loved to be needed and was a sucker for a bit of praise. She would talk about bathing newborns, cradling the precious head in the palm of her hand, the first cry, her eyes wide and still marvelling at the miracle and beauty of new life.

The African ward sister was worshipped. It was 'Sister Abuoa this' and 'Sister Abuoa that'. We imagined a stern, unsmiling headmistress, all tough compassion, cutting through protocol when needed, sticking to it like a barnacle when it was the right thing to do. It was my mother's deepest regret that she would never be a proper midwife, with the whole of the birth, the child's and the mother's life under her care and control.

She always took it hard when babies were born dead. She had lost a couple herself between the two of us, siblings we never knew about so never missed. Sister Abuoa would send my mother off to the canteen or staff room after the terrible job was done, and just once she sent her home.

We were sitting in the kitchen with tea and toast when she opened the door. She had a big bag of shopping so we took that as the reason she slumped in a chair, exhausted and red-eyed. We paid her no attention.

'I've been walking around since four o'clock in the morning,' she said. We looked at one another. She spoke from somewhere far away, still in her thick coat and woolly hat, her square hands knitted together on her lap. 'I had to wait for the market to open.'

She got up suddenly and looked around the kitchen like she was seeing it for the first time. 'I'll start the dinner,' she said. 'Then I'm off for a sleep. You two can tidy this kitchen.'

We sat where we were, me reading an old magazine that told me how best to pluck my eyebrows and my sister half asleep, unfurling her black hair from big sponge rollers.

My mother threw her hat and coat on a chair, tipped the vegetables out of the bag and started peeling the potatoes, cleaving them in half and dropping them into a saucepan of cold water. It was our job to put the dinner on late afternoon so it was ready when she woke up, before she went back to work.

I don't know what made me look up. She gasped, I think. She was staring at the chicken she had bought. She turned it over, front then back, and put one hand tenderly on its breast. She stayed like that for a moment. Then she stuck her other hand in, wrist-deep, right up to the neck, and grabbed something. I heard the suction as she pulled it out. It was the heart. The tiny heart.

She turned and looked at me. 'We did everything we could,' she said. 'No one could have done more,' she whispered.

'Yes, Mum,' I said.

'You never forget,' she said. 'That's part of the job. Being there and remembering.'

She rinsed her hands and wiped her face with the tea towel. She wrapped the chicken back in the newspaper and put it in the fridge.

'Fish and chips tonight,' she smiled. 'For a treat.'

MARK GATISS

Nothing about this is easy to write. My mother – a little dot of funny Anglo-Irishness with a lovely singing voice (after a glass of Carlsberg) – was diagnosed with lung cancer at the age of 71. I can remember with horrible clarity the day she came back from the doctor, having been told there was no hope – my sister, Jill, who had accompanied her, giving a minute shake of the head to us as they came through the door. Mam was remarkably stoic. More so than my poor Dad, who found it difficult to discuss the reality of it all.

I suppose we all wanted things to be normal for as long as possible. It's only in such circumstances that you appreciate what 'normal' is. And that it's the small things – laughter, talking, togetherness – that bind us all. Throughout this time, brilliant Macmillan nurses visited and helped Mam through her chemo. It's impossible to speak too highly of their innate kindness and good humour. It made a huge difference, and the yellow glow of that daffodil emblem still makes me smile, even though the times were so dark.

Jill, who had worked as an NHS nurse since the early 1980s, had by this time become a district nurse and was, to our great relief, permitted to look after Mam at home. Thus, the last weeks of Mam's life were spent there, surrounded by the people who loved her. It's a strange contradiction that those times, sitting around the bed and swapping old stories, taking turns to make tea and sandwiches, are very happy memories.

My sister brought all her years of professional experience to looking after Mam, even though it was heartbreaking to endure, and when the end came it was a very good death. I can remember my sister holding Mam's hand and then bowing her head to rest on the bed as Mam slipped away, like something from a medieval etching. It's the sort of thing you never imagine you'll witness or be part of.

A few years later, my sister found a lump in her breast. With cruel irony we realised the cancer must have been growing inside her all the time she was looking after Mam. Jill approached the end with similar stoicism. Not as a 'battle' (she knew exactly how long she had left), but as a process. A process helped immeasurably by being surrounded by her fellow NHS nurses who were also some of her oldest friends. Back on her old ward, this time as a patient, she was literally prescriptive about the drugs she needed! The atmosphere of good humour and gentle, practical care the staff engendered through the long days and nights that followed was unforgettable. No one likes hospitals. But to see that staff, that whole building, doing its best in the face of creaking infrastructure and underfunding to create a quiet, dignified end will stay with me always.

My brother-in-law, Dave, was devastated, of course. Always a source of quiet fun and kindness, he managed well, taking over sole parenting of their two boys but always feeling Jill's absence. Who can say what the stress of this hammer blow did to him? At any rate, he suffered a massive stroke a few years after Jill's death. Yet again, the NHS was there for him, through night vigils and painful visits. Yet again we found ourselves grouped around the sickbed, listening to every laboured breath, holding hands, bound together in our pain as he passed.

Nothing about this is easy to write. Except that knowing there is a system of care, a net to catch us when we fall is one of this country's greatest success stories. That it has taken this terrible emergency to demonstrate this fact is telling and sad, but it's also an opportunity.

We simply couldn't survive without the NHS.

It is the best of us.

FRANKIE BOYLE

This is a column I wrote for the *Guardian* a few years ago about the way that then-Minister for Health Jeremy Hunt was presiding over the managed decline of the NHS. In retrospect, Hunt seems at least to be an energised character compared to the current incumbent, Matt Hancock, who always looks like he's still in the friend zone with his own wife. Yet I think it's important to look back, because a key thing we should learn from this pandemic is the way that the gradual, deliberate underfunding of the NHS has contributed to the current disaster.

One of the worst things for doctors must be that, after seven years of study and then another decade of continuing professional exams, patients come in telling them they're wrong after spending twenty minutes on Google. So imagine how doctors must feel about Jeremy Hunt, who hasn't even had the decency to go on the internet.

Consider how desperate these doctors are: so desperate

that they want to talk to Jeremy Hunt. Even Hunt's wife would rather spend a sleepless seventy-two hours gazing into a cracked-open ribcage than talk to him. Hunt won't speak to the doctors, even though doctors are the people who know how hospitals work. Hunt's only other job was founding HotCourses magazine: his areas of expertise are how to bullet point a list and make dog grooming look like a viable career change.

Of course, the strikers are saying this is about safety not pay, as expecting to be paid a decent wage for a difficult and highly skilled job is now considered selfish. Surely expecting someone to work for free while people all around them are dying of cancer is only appropriate for the early stages of *The X Factor*. Sadly, Tories don't understand why someone would stay in a job for decency and love, when their mother was never around long enough to find out what language the nanny spoke.

The fact that Hunt wrote a book about how to dismantle the NHS makes him feel like a broad stroke in a heavy-handed satire. Even the name Jeremy Hunt is so redolent of upper-class brutality that it feels like he belongs in one of those Martin Amis books where working-class people are called things like Dave Rubbish and Billy Darts. (No shade, Martin – I'm just a joke writer; I envy real writers, their metaphors and similes taking off into the imagination sky like big birds or something.) Indeed, he's so overtly ridiculous that Jeremy Hunt might be best thought of as a sort of rodeo clown, put there simply to distract the enraged public.

I sympathise a little with Hunt – he was born into military aristocracy, a cousin of the Queen, went to Charterhouse, then Oxford, then into PR; trying to get him to understand the life of an overworked student nurse is like trying to get an Amazonian tree frog to understand the plot of *Blade Runner*. Hunt doesn't understand the need to pay doctors – he's part of a ruling class that would happily scalpel out a prostitute's womb as part of a stag weekend. He comes from a section of society that doesn't understand that the desire to cut someone open and rearrange their internal organs can come from a desire to help others, and not just hereditary syphilis.

The government believes that death rates are going up because doctors are lazy, rather than because we've started making disabled people work on building sites. Indeed, death rates in the NHS are going up, albeit largely among doctors. From the steel mines where child slaves gather surgical steel all the way up to senior doctors working thirty-six hours on no sleep, the healthiest people in the NHS are actually the patients. This is before we get to plans for bursaries to be withdrawn from student nurses, so that we're now essentially asking them to pay to work. Student nurses are essential; not only are they a vital part of staffing hospitals, they're usually the only people there able to smile at a dying patient without screaming 'TAKE ME WITH YOU!'

The real reason more people die at a weekend is that British people have to be really sick to stay in hospital over a weekend, as hospitals tend not to have a bar. We have a fairly low proportion of people who are doctors and don't plan to invest in training any more, and are too racist to import them. So we're shuffling around the doctors we do have to the weekend, when not a lot of people are admitted, from the week, when it's busy. This is part of a conscious strategy to run the service down to a point where privatisation can be sold to the public as a way of improving things.

Naturally, things won't actually be improved, they'll be sold to something like Virgin Health. Virgin can't get the toilets to work on a train from Glasgow to London, so it's time we encouraged them to branch out into something less challenging, like transplant surgery. With the rate the NHS is being privatised, it won't be long before consultations will be by Skype with a doctor in Bangalore. Thank God we're raising a generation who are so comfortable getting naked online. 'I'm afraid it looks like you've had a stroke. No, my mistake, you're just buffering.'

When I was little, I was in hospital for a few days. The boy in the next bed was an officious little guy who took me on a tour of the ward. He'd sort of appointed himself as an auxiliary nurse and would help out around the place, tidying up the toys in the playroom and giving all the nurses a very formal 'Good morning', which always made me laugh. I got jelly and ice cream one evening (I'd had my tonsils out) and they brought him some too. Afterwards, he threw his spoon triumphantly into his plate and laughed till there were tears in his eyes. Then he tidied up and took our plates back to the trolley. What he meant by all this (we'd sit up at night talking and waiting for trains to go by in the distance) is that this was the first place he'd known any real kindness, and he wished to return it. For most of us, it will be the last place we know kindness. How sad that we have allowed it to fall into the hands of dreadful people who know no compassion, not even for themselves.

JACKIE KAY

Mask

SAGE, SARS, PPE
NHS, BAME
Abbreviations we understand;
But not lives shortened
By too few goggles, gloves, gowns,
Johnson doing a ring around,
Carers compelled to wear bin bags.
Doctors decked out in out-of-date masks. Time lags.

But Death has no abbreviation. No hiding place.
Wash your hands! Don't touch your face!
Don't listen. Close your ears. Don't test, test, test.
We will come to know your names –
Our carers, nurses, doctors, our NHS.
And we will mask our heads in shame.

STEWART LEE

About fifteen years ago, when my then-fiancée had just moved in with me, I had really bad diarrhoea. I'd had diarrhoea all the time really for about twenty years but I hadn't told her about that as it was exactly the sort of thing she wouldn't like, and I was getting to the age where I should probably get married.

I lived in a small, one-bedroom flat above an estate agent's in north London and I went to the toilet as usual with the diarrhoea and it stank the whole flat out. In fact, it stank everything up so bad on this occasion that the woman who ran the estate agent's downstairs actually rang the doorbell and asked my fiancée if there was some problem with the pipes.

My fiancée told her there was no problem with the pipes but that I had done a really bad diarrhoea that had stunk everything up and that she was getting fed up of it. The estate agent and my fiancée both agreed that I should go to the doctor soon, as it was awful and I had admitted it had been going on for ages. I had never been especially ill in my thirty-seven years of life ever, apart from the diarrhoea, which didn't really bother me, so I was not registered with any doctor's surgery.

And also, when I was born it was in a secret place and I was sent to a charity home with a different name that was then forgotten, so there may have been no record of me existing to the NHS. My fiancée got me seen as a guest patient at her nearby GP.

I walked slowly along the road, past the pubs and ice-cream parlours and Indian restaurants where I had spent most of my twenties and thirties, and into the doctor's by the park. The nice receptionist took one look at me and said, in a friendly way, 'Are you the diarrhoea? You probably don't need to see a doctor. Just read this leaflet.'

I sat in the waiting room, which was bright and had a kind of sculpture of some human intestines on the wall and a place where children could play in a pretend kitchen made of wood. There were also some old *Mojo* magazines and local newspapers to read. It was nicer than my home. I looked at the leaflet. It said I should avoid fizzy lager and white wine, dairy products and spicy foods.

I asked the lady how much the leaflet cost and she said it was free from the NHS. I could not believe it. So I said thanks and goodbye and I went away and did all that the leaflet suggested and I have been fine ever since, although I have put on a lot of weight by not having constant diarrhoea. The NHS is amazing though, really, as I have not had diarrhoea since it gave me that leaflet.

I know the NHS is being incredible at the moment in the virus, and I wanted to help with this book, but I haven't got any personal experiences, apart from the diarrhoea leaflet one, and I don't really have any friends or family to speak of, so I can't use their stories.

I saw a programme on TV where a man's leg came off and they found it on the road later and made it go back on, but that isn't a personal story. It might have been his arm actually. Anyway, it was good.

I am sorry I have not been able to be more help as the NHS people really are doing an amazing job. I understand Dawn French is submitting something. I met her in Cornwall last year, but while she was talking to me I tripped over this sort of free-standing advertising banner and it fell down so it was all a bit awkward and then she went away. Anyway, I am sure Dawn's contribution will more than make up for this one being not that great.

If there is a fee for this I am happy for half of it to go to charity. If not, please send to my agent to deal with and thanks for asking me.

E L JAMES

My Earliest Memory

I can picture my parents standing in a doorway, holding a large inflatable rabbit. From memory, he was green and pink and humongous – probably bigger than I was at the time. He was an odd choice for a sick child and, looking back, I wonder what the hell my parents were thinking. My first language was my Chilean mother's tongue, so I named him 'Gringo'. (I know, I know!) To a wee girl of two, this was just a funny and familiar word, the epithet my mother often used for my Scottish father, fondly … and occasionally not so fondly.

This memory might be the remnants of a nightmare – and I imagine it was, for my parents. Or perhaps Gringo was a product of my fevered two-year-old mind, a thought just occurring to me now as I write this. But what's even weirder is that I see Mum and Dad as their forty-year-old selves, not the overwrought twenty-somethings they would have been at the time. It's sobering how one's mind and the passage of time can play such tricks. Sadly, I can't ask either of them for more details; my father is no longer with us and my mother suffers from severe Alzheimer's. Her memory is a distant acquaintance who rarely visits, and never writes thank-you letters.

I know it was 1965 and I was in Stoke Mandeville Hospital in Buckinghamshire. When I was first admitted, the doctors had suspected that I had meningitis. Fortunately, it wasn't that, but what I did have was a critical case of double pneumonia. I was on the edge of death, my temperature so high they thought they would lose me. Back in the day, during a more lucid era, my mother told me that it was the only time she'd ever seen my father pray; he was a confirmed atheist – so much so that the priest from Our Lady of Lourdes Catholic Church in Acton, who married my parents in 1962, begged my mother to reconsider her choice of husband.

My mum told me – again, I can't check – that I was given a new drug that reduced my temperature so quickly I had to be wrapped in tin foil, like a small basted joint, to warm me back up.

I recovered fully, saved by the staff of the NHS, to whom I remain eternally grateful. However, my near-death experience from a respiratory disease decades ago may explain my very real fear of this horrible virus now affecting so many of us.

Thank you to the brave men and women who work for our beloved NHS. We all owe you so, so much. I hope that, in time, we as a nation can repay this debt of gratitude with better appreciation, better facilities and better pay for everything you do.

Much love.

MARINA HYDE

My first labour was 59 hours long and I watched three episodes of *I'm a Celebrity* . . . live during it. To put that into some kind of historic perspective, my waters broke at the moment Gillian McKeith returned to camp with nought stars, leaving the celebrities on basic rations for the foreseeable future, and me suffering contractions every ten minutes. Also, apparently, for the foreseeable future. A situation handled caringly, and eventually chemically, by the sainted Queen Charlotte's Hospital in west London.

Returning twenty-one months later to the same place, I regarded my next son's birth, which took place over eighteen hours, as a mere bagatelle. Twenty-four hours later, we took him home, and twelve hours after that we took him to the A&E department of St Mary's Hospital, Paddington. Our tour of London hospitals would, over the next few weeks, take in both the Chelsea and Westminster and Great Ormond Street. To say I love and worship them all would be an understatement; at the time, I would have died for them if it meant my baby could be saved.

What was wrong with him? Forgive the foray into deep clinicalese, but it was eventually established that his insides were all twisted up. It is the only time since the internet has been in our lives that I didn't look something up to find out more about it. I simply listened, instead, to the cavalcade of angels who worked in neonatal emergencies at those hospitals. And I thoroughly recommend paying attention to people who say things like: 'We really think it would be best for everyone if you weren't in the room for the lumbar puncture.' Also, don't Google 'neonatal lumbar puncture'. I still haven't.

I say that his diagnosis 'was eventually established'; in fact, it took only two days. Those two days weren't simply the longest of my life – I firmly believe entire civilisations rose and fell in them, while ice ages swept in and out. I spent those epochs asking everyone from doctors to nurses to cleaners to random people in corridors whether my baby was going to die and was only answered with what felt like an increasingly tentative, 'Look, he's in the best place . . .'

The surgeon, Will Sherwood, at the Chelsea and Westminster, was the first to say 'No'. No, he wasn't going to die, because he was going to operate on him later that day. I still don't know the precise details of what went on during those hours because I was too gibberingly fearful to ask in full.

When we were out of the worst woods, I returned to

Google. My search terms were limited to 'possible to fall in actual love with surgeon' and 'possible for husband to also fall in actual love with surgeon'. I obviously resolved to marry Will to one of my sisters and was hugely put out during one of the night vigils with the brilliant nurses when one of them confided that he was spoken for. 'What a surprise,' I hissed over the nest of tubes. 'He's young, good-looking and his job is SAVING BABIES. I literally can't believe he's not single.' (You'll note it wasn't all surgical wins for us: my sarcasmectomy was unsuccessful.)

Nearly eight years on, I can still remember the names of the nurses who helped us, the cleaners who cheered me after night-time vigils, every extraordinary surgeon and consultant – all the people who did so much for us that when you finally leave you honestly can't believe that you aren't simply giving them everything you own to say thank you. But that, of course, is the immense and wondrous beauty of our NHS. You just say thank you.

When I told my son about this book, and that I was going to choose this – the jewel in the crown of our myriad family NHS stories – he wondered why I wasn't going to do the one about when he used the lamp as a light sabre and St Mary's spent ages picking the glass and burnt skin out of his arm while he chatted to them in his stormtrooper outfit. Two reasons, I said. Firstly, I'm still massively annoyed about that use of the lamp. And secondly, the origin story that enabled him to disport himself with decorative household electrics is – let's face it – kind of the bigger one.

Interjection from the elder brother: 'It's so unfair that he gets to be in the NHS book! I want to be in the NHS book!' Thoughtful pause. 'If I cut off my leg, can I be in the NHS book?' Interjection from the younger sister, after an even more thoughtful pause: 'If I cut off his leg for him, can I be in the NHS book?'

Thank you, NHS, for performing the daily miracles without which there would be no one to have heartwarmingly violent family negotiations like this. Thank you, NHS, for each of my most beloved little negotiators, all of whom will be making further NHS stories with you their entire lives. But hey, this particular one is his story, and it's all thanks to you. So thank you, NHS, for my Otto.

ROBERT WEBB

Dear NHS,

Let's face it, most times we encounter the NHS we don't tend to be in the sunniest of moods. It's like a friend who reliably turns up on all of your shittest days. You feel like going, 'Oh, YOU again! What is it with you? What is this weird interest you have with my health, you freakster?' But then, very quickly, that tends to become, 'Oh, I see. Yes, that does feel better. Will it . . . ? Oh, so I should just . . . ? OK, well thanks. That's very nice of you. Thanks very much. And thanks for always being there.'

The NHS really is unique among public bodies. We don't tend to feel the same level of affection for, say, the Citizen's Advice Bureau or the Department for Work and Pensions. I think the difference is partly to do with the people. Don't get me wrong – I'm sure both the CAB and the DWP are stuffed to the rafters with dedicated professionals who are all wonderful, as well as sexually attractive, human beings. But I imagine the work involved in those places and I (rather lazily, no doubt) think, 'Yeah, I could probably do that.'

But then I imagine working in a hospital. And at that point I go a bit quiet. The work done on a daily basis by doctors, nurses and all the other frontline staff is quite beyond the likes of me. I like to think I'm not without compassion, but to dedicate your whole life to looking after people when they're sick . . . that's something else.

There are also the non-grumpy, non-miserable moments too. Thank you, NHS. Thank you for safely delivering our two daughters. Mainly, I thank their mother. But we had help.

IAN RANKIN

I was a timid, risk-averse kid. Not sporty; pretty terrified of heights. Happier with my comics and music in my bedroom than on the sports field or a racing bike. So I never cut myself badly or suffered a concussion or broke any bones. I was also healthy (or do I mean lucky?) – no tonsils or appendix requiring removal – meaning I was no great burden to any NHS hospital. Mind you, I grew up in the halcyon days of home GP visits and I did catch every childhood ailment going: mumps, measles, chickenpox, you name it.

My lucky streak ended, however, in 1979, the year I turned nineteen. My mother died of cancer that June, having been increasingly ill for the previous eight months. I was still processing that when, in August, I had to take time off work (a chicken hatchery near my home in Fife) because I was suffering blinding headaches. The initial diagnosis was a migraine but the prescribed tablets didn't help in the least. I ended up unable to get out of bed and, after my father's traditional cure (a vinegar-soaked cloth on the forehead) proved ineffectual, a home visit by our GP led to a new diagnosis of meningitis.

I was rushed into hospital and given a lumbar puncture. Curled in the foetal position, I didn't see the liquid spurt halfway across the room from my spinal column but the attendants assured me it was a spectacular display. (I dare say these days a phone might have captured the moment for internet posterity.) Along the way, I was also found to be giving a home to some form of salmonella bacteria. I was confined to a room of my own with a TV and radio, so once the various procedures had been dealt with I could relax and wait for the all-clear.

The nursing staff were great, lingering especially long if a good film was playing of an evening (Clint Eastwood was a favourite). I also recorded in my diary that my consultant would breeze in at daily intervals, do the crossword in my newspaper, then breeze out again, leaving me one less means of passing the time.

If the hospital was understaffed or under-resourced, everyone did a good job of hiding it from the patient. My father is long dead now, but thinking back, I do wonder at the stress and worry he must have felt in those few brief months. Me, I was young and bored, mostly. A friend sent a Snoopy book and Tolkien's *The Hobbit*. Both sustained me in a way the Chaucer I was supposed to be studying (I was in the hinterland between first and second years at university) could not.

I was not to make much use of the NHS in the fifteen years that followed. However, in 1994 my son was born with special needs. He was our second child and at first we were fairly relaxed. We had read the parenting books and knew that infants developed at different rates. Finally, around the age of nine months, we had a diagnosis of Angelman Syndrome. Children's hospitals and paediatric specialists became the norm for a time as he was tested and tested and tested again. There were seizures to be controlled and scoliosis (curvature of the spine) was yet one more problem to be factored in. The doctors and nurses got to know us and we got to know them.

Our son is in his mid-twenties now. He grew out of the seizures and the scoliosis did not worsen appreciably. But he cannot walk or feed himself or communicate, so whenever he is under the weather there are always more tests and procedures and tense waits for answers. Nor is he always the most accommodating client. It takes three of us to cajole and control him when his teeth need to have their regular examination, and dental procedures beyond a check-up require a hospital visit and general anaesthetic. It doesn't help that he has arms like an octopus and an uncanny ability to notice and then attach his vice-like grip to the one thing in the room you wish he wouldn't.

The people I've encountered in every area of the NHS have been professional, caring, empathetic and patient. The GP of *Dr Finlay's Casebook* may be a thing of the past, but our GP practice in Edinburgh has always been there for us. The virus of 2020 has reminded us how much we depend on these professionals – from the hospital porters and cleaners right up to the most senior consultants and surgeons. They are ordinary people doing often extraordinary things. Those vinegar compresses – however well-meant – were not going to save my life and all the parental love in the world was not going to ameliorate my son's condition. But the NHS did, and the NHS do.

And I know I'm by no means alone in being indebted to them.

LAUREN CHILD

Almost exactly a month after my sixteenth birthday, my appendix popped. It didn't happen completely without warning, it took approximately seventy-two hours. Beginning on the last day of the Christmas holidays when it was snowing – proper snow, the sort that stops the traffic.

I'd spent the day with friends, wondering at the strangeness of our now silent town which appeared 'rubbed out' – all its edges gone. I was aware that something strange was also happening to me. I thought it was the cold that was causing this odd feeling. By the next morning, it had become an ache that became a pain which grew into something akin to a wolf gnawing at my insides.

I didn't make a big fuss about it, not because it wasn't agony but because I wasn't entirely sure whether *saying* it was agony would be considered 'making a fuss'. This was my first reason: in my family, you needed to be mortally injured *not* to be 'making a fuss' and how could you claim you were *in agony* without 'making a fuss'?

Secondly, I was lying about the location of the pain. If I told the truth, I feared my mother would enquire if it might be my period? – this word always uttered in a whisper on the rare occasions it had to be spoken – being eaten by a wolf was certainly preferable to this.

Thirdly, I didn't want my parents to call a doctor. I had a deep fear of doctors – all those embarrassing questions – and I could think of little worse than one coming to the house.

In the end, my mother did call a doctor, who immediately called an ambulance. This was *very* embarrassing because the ambulance arrived at exactly the same time as everyone from my school was walking past my house. While my parents were awkwardly explaining their delay in phoning, I was considering the tragedy of just how many of my classmates might have caught a glimpse of me in my sweat-soaked pyjamas.

The roads were closed due to the ongoing blizzard and they couldn't get me to the big hospital in Swindon, so I was taken a mile up the road to Savernake, the cottage hospital on the edge of town. This was a place very familiar to me; my sisters and I had made various visits over the years to have squashed fingers stitched, greet newborn babies and watch our grandmother fade away.

Even as I was wheeled to theatre, being at death's door did not assuage my embarrassment. I had not washed for three days and my hair was tangled and damply stuck to my face. I was very conscious of this. Yet what I remember most clearly was how nice everyone was to me, incredibly nice. They made no mention of my unclean state. There was no eye rolling, no holding of noses, one nurse even held my hand. No judgement, just kind words and a nurse who held my hand.

When I came round from the operation, I appeared to be on a geriatric ward. The patients were all extremely ancient, although thinking about it some of them may only have been ten years older than I am now. I didn't mind a bit because the 'old ladies' were sweet to me and funny to listen to, and I didn't have the energy to talk. They were always trying to get me to eat but I didn't have the energy for that either – not even the Yorkie bar my sister had placed on the shelf beside my bed.

This bar was spotted by my nurse, who promptly picked it up, broke off the end and ate it. 'I love the ends,' she said.

She came back the next day and ate its other end. 'It's the ends I love,' she explained, 'I'm not interested in the rest.'

What I did have the energy for was walking to the toilet and no amount of imploring or scolding was going to stop me.

The kind Yorkie nurse seemed to understand a sixteen-year-old girl who was embarrassed by bed baths and bed pans and too-personal questions.

Nurse: 'Have you had a movement?'

Me: 'Pardon?'

Nurse: 'A number two?'

Me: 'Sorry?'

Nurse: 'A poo?'

Me: silence.

The nurse told me I was lucky. She said that I'd come in 'with just two hours to live'. She said my survival was down to the surgeon, 'He's the best in the area.'

She was right, I was lucky. Lucky for the snow which meant I was taken to our small-town hospital, no trouble for friends to visit. Lucky that we *had* a small-town hospital. Lucky that the best surgeon in the area lived in our town. Lucky to have a nurse who understood me. And lucky for my fellow patients who made me laugh.

The whole experience changed my life, and not simply by saving it.

ALEX BROOKER

People always say to me, it must have been awful having all those operations as a kid. But to be honest, it wasn't really, not at all.

And that's because of the NHS and Great Ormond Street Hospital. For me and my family, Great Ormond Street became like a second home and like a second family. In a way, it's like a hospital version of Disneyland.

I looked forward to going, if anything. All the bright decoration, the lights, the statue of Peter Pan outside, the Super Nintendos in the wards, the waiting rooms – and the staff, of course.

All I remember are smiling, reassuring faces, and whether you're a kid about to have their umpteenth operation or their first they mean everything.

Of course, there were times I was sad and even scared but there was always someone on hand, whether that was a nurse, a play co-ordinator or another amazing member of staff to make me feel better.

I used to lie in bed and listen to all the noise and hustle and bustle of central London outside and would imagine myself out there living the exciting lives of those people. But Great Ormond Street and the NHS made that a possibility.

They amputated my foot to get me up on my feet (one now being metal); they gave me an opposable thumb on my left hand so I could write, drive a car and lead an independent life. They tried to make sure it didn't leave a big scar, because a kid getting conscious about his hands asked them to.

People always say to me, it must have been awful going through all that pain. But twenty-five-odd years on, I'm not writing about pain. I'm writing about how wonderful the NHS and Great Ormond Street Hospital have been to me. Those are the memories I'm left with.

So I'd just like to say thank you for looking after me, and for helping me to lead a life even more exciting than the one I was imagining when I was lying in that bed on Clarence Ward.

KATE ATKINSON

Dear NHS,

Thank you for the drugs. I don't know if you can get them any more – probably not – but they were good. Trust me, I'm not a doctor.

I'm talking here with particular reference to childbirth, how it used to be back in the Stone Age or, to be more precise, 1974, because the older I get the more my past feels like living history. So this is how it was then.

In my defence here, I'd like to point out that I was very young in 1974 and had no idea about anything, but particularly not babies. I had never held a baby; I'm not sure I'd ever even seen a baby. We lived in a cottage on a farm in Fife with one dog, two goats, five cats, six chickens and quite a lot of rats. I swear to God that I genuinely thought having a baby would just be like getting another kitten. I didn't even have a washing machine, it was my mother who pointed out to me that I probably needed one and it was my mother who bought it as, I'm ashamed to admit, she bought nearly everything the baby needed. Those were my careless days.

My mother had given birth to me in a private nursing home in order to draw a discreet veil over my illegitimate status (it was 1951). She spent what was then a standard three weeks in the home, mostly in bed. It was the middle of a very cold winter and she came out with chilblains and a conviction that I had been accidentally swapped with another woman's baby. 'I'm sure I know who your real mother was,' she occasionally said to me when I was growing up. I always suspected I had the wrong mother so I guess that made us evens.

Maternity care had evolved by 1974, but perhaps not much. I had one antenatal check-up and one scan, that was it. Nobody took their scans home and cooed over them in 1974; nobody considered trying to find out the sex of a baby. Nobody was obsessed with natural childbirth. Babies were just something that *happened* to people. I did have a book (of course I had a book). It was titled something like *Having a Baby* and I had been about to read it when I went into labour three weeks early.

The most troublesome thing at the time was that we lived on the other side of a very wide river and the only way to cross it was by a toll bridge, so we had to spend some precious time scrabbling around looking for change. (We were not just young, we were also poor.) Looking back, I'm pretty sure that they would have let us over without paying, but you never know. She could have been the first baby born with 'Tay Bridge' for 'place of birth' on her birth certificate. Because she was a girl – something that I had been one hundred per cent convinced of from the very beginning. (I'm pretty good at predicting the sex of a baby, one of my many wasted minor talents.)

I know what I was wearing when we eventually arrived at the hospital – a long, black, velvet cloak. It drew some odd looks. I had been married in it too. Everyone got married young in 1974. I had two favoured items of clothing during the later stages of pregnancy – both long cotton nightdresses made for my grandmother when she was visiting her war-bride daughter in Canada in 1948. We weren't hippies, just students who didn't know any better.

It was the middle of the night by the time we crossed the bridge. A solitary midwife was in attendance when we arrived at the hospital; she was Irish, middle-aged, red-haired. I can see her now. Don't worry, we're getting to the drugs.

Pethidine! A strong opioid. No consultation, the midwife just whacked it in. I'm looking at it now online – I have four contraindications to taking it. *Four*. It was great, like Valium, but better. It was four in the morning, the time of least resistance. And magically, after hardly a twinge of pain, there was a baby. A baby! I think I was still half expecting a litter of kittens.

The midwife, the most focused woman I've ever encountered, let her brain go off shift immediately. 'It's a boy,' she said. I shot up, despite the strong opioid, and said,

'No, it's not.'

'Oh, you're right, it isn't,' she agreed and disappeared, never to be seen again. She was replaced by a doctor, the first and last time that I saw one that night. A man, needless to say. (Again – 1974.) He arrived just in time for the pethidine to have worn off. He 'sewed me up', anyway. I didn't even realise I'd had an episiotomy until that moment, thanks again to the pethidine. Being sewn up without an anaesthetic ten minutes after you've given birth is bad! Very, very bad.

But then it got good again because, in those days, they kept you in hospital for a whole week (no MRSA to fear then). They didn't care whether you got out of bed or not and brought you endless milky drinks and biscuits and they took the baby away every night and did God knows what with it. We mothers certainly didn't know because we were given a very strong sleeping tablet every night. Amazing! It was the only time I've ever had sleeping tablets and it was lovely. You just floated off until you were woken with tea and toast the next morning and handed a placid baby that had obviously been pumped full of formula all night.

I was on an old-fashioned, fourteen-bed Nightingale ward and I was the only one breastfeeding. That's 1974 for you. You can imagine it's quite difficult to breastfeed when you've had pethidine (breastfeeding being one of those contraindications), as well as sleeping tablets and said baby is already being clandestinely bottle-fed. If only

I'd had time to read that book I might have known these things, might have put up a weak cry of protest. But, dear God, those were the seven best nights' sleep I've ever had. So, thank you, NHS.

Second time around, 1981, was a different world. In and out in twenty-four hours, no drugs, just yoga breathing, no episiotomy – I seem to remember shrieking, 'Don't cut me!' to someone. And I didn't need a book because I'd had that first baby to practise on. Still, 1974, I salute you.

Kate x

P.S. Thanks also for the penicillin that probably saved my life in 1958 and then again in 1972. And probably 1999 as well.

CANDICE CARTY-WILLIAMS

Dear NHS,

Physically, I am very hardy. I'm not one of those people Natasha Bedingfield sings of in her UK top 15 hit 'I Bruise Easily'. I've never broken a bone and apparently I have a good enough number of white blood cells that if and when I get ill, I recover fairly quickly. Where I am not hardy, though, is my brain. I am one of those fragile but resilient people who found it easier to take pain in, bury it and keep on pushing. This caught up with me a few years ago, though, when I was twenty-three.

Following the loss of my best friend, my world started to spin off its axis. Panic attacks came and depression followed soon after. For roughly two years, I was stuck in my bedroom in a deep pit of despair, overwhelming past memories coming at me left, right and centre. It was debilitating and I was afraid. This all came neatly packaged in the form of a phobia. This phobia is where I directed all of my energy. I spent months researching this phobia,

I spent months feeding it and accepting that it had ruined not just my life at present but my life to come. But, on one sunny day, I decided that it didn't have to always be like this. I was too weary and too beaten down by everything, but a glimmer of hope in the form of the sunshine through my window told me that I should try to sort out things in my head. I geared myself up, went to the doctor's down the road (a five-minute walk that felt like a five-year one) and I told them that, in my research, I'd seen that cognitive behavioural therapy (CBT) might help me.

The doctor listened and referred me on. I had an assessment (painful), told the person on the phone about my phobia in great detail (especially painful) and was told that the waiting list for CBT meant that I'd be seen in 'around six weeks to two months' (the worst pain). A few days later, I called them back. I told them, with no detail spared, that I couldn't go on like this and that I certainly couldn't wait two months. There was a lot of crying (I think on both sides) and the next week, I made my way to the Maudsley Hospital's Centre for Anxiety Disorders and Trauma in Camberwell. I'll always remember my bus journey there. I felt terrified, but I felt hopeful.

Over the next few weeks, I worked with my therapist to sort my phobia out. To place where it had come from, to understand it fully and to undo the ways it had embedded and twisted itself in my brain. It was the toughest, most testing, most *exhausting* three months of my life. The middle was particularly hard. For one session, a friend had to do the journey with me just to make sure that I went. At the end, though I wasn't miraculously healed (that isn't how phobias work, you see), I was able to use the tools I'd learned to change my life as it was and move forward.

Years on, the phobia can still rear its very ugly and especially unwelcome head, but I force myself to put into place all that I worked on six years ago. I was, and still am, so grateful for what the NHS gave me. Not just the hope, not just the learning, but it gave my life back to me and it made me both weaker and stronger. Weaker in that I learned to accept that I didn't always have to be strong, and stronger for knowing that there's strength in vulnerability.

And for that, I thank you.

JOSH WIDDICOMBE

Last year, when I was in hospital, the anaesthetist told me that he thought they should legalise heroin. I'd never been under anaesthetic before and I wondered if this was part of the process, like counting to ten until you have passed out. 'Good news, we know he's under; he stopped talking around the point he agreed that VAT on drugs could help the NHS.'

I don't have any strong views on the heroin question, not the kind of views I would share with strangers injecting me with morphine anyway, so I nodded and smiled and tried to move the conversation on to Arsene Wenger, like I do in the back of a taxi. And then passed out.

Support the NHS.

VICTORIA DERBYSHIRE

I'd just said 'I love you' to a Greek radiologist I met only three hours earlier. It wasn't a date.

Aged forty-six, I had a shock breast cancer diagnosis. It was a profound moment in mine and my family's life and turned everything upside down. I'd only ever been in hospital to give birth. I didn't even know my local hospital, Ashford and St Peter's in Surrey, had a breast care clinic.

Demitrios Tzias was one of a team of NHS specialists treating me. He had a low, reassuring voice (he'd have sounded terrific on the radio) as he explained the ultrasound he was carrying out on my left breast was to check if the cancer had spread. I was having my right breast removed because that definitely was cancerous. My partner Mark and our two boys, aged eight and eleven, were with me for support and they all engaged in football chat – remember when Greece won the Euros? It occurred to me to mention to Demitrios the couple of dizzy spells I'd experienced recently and the room suddenly went silent. He asked the boys to wait outside and Demetrios looked me straight in the eye and calmly said, 'In that case, I think you should have a full CT scan which will check your body and your brain.'

It took one heartbeat for us to absorb that information. 'Oh god,' I said. It was the fact that Demitrios had specifically mentioned the brain.

There was more silence as the significance of what he'd said hung there. Then the questions came: if there was apparently no cancerous tissue in my lymph nodes (which there wasn't), how was it possible that the cancer could have bypassed them and travelled to my brain? It was possible, he explained; unusual, but possible.

My heart started to crack and I panicked inside. I tried to process what was happening and couldn't. Mark gently held both my arms and said something reassuring about the odds of it having travelled to my brain being unbelievably low.

Within half an hour, Demitrios had arranged for me to have the scan and I lay on a white bed that rolled back into the CT tube. It was claustrophobic but that was the least of my worries. It took only ten minutes and Demitrios told us before we left that he'd try to let me know that afternoon what the result was.

Driving the short distance home it felt like I was suffocating. We pretended everything was normal to the boys. As soon as we arrived back, they went outside to play football and Mark and I sat opposite each other at the kitchen table. We had to talk about it there and then, there was no tip-toeing around it. I took a deep breath and told Mark calmly he might have to bring up the boys on his own. He immediately said that wasn't going to happen, it couldn't happen. He looked forlorn and I didn't blame him in the slightest, but it was surreal how utterly focused I was on the practicalities – how much equity there was in the house, how it would be possible to pay off the mortgage and downsize if I wasn't around any more.

Forty-five minutes later, as I lay on my bed staring at the ceiling, my mobile rang. It was Demitrios. Shit, this was it. I could barely breathe. I didn't speak, simply listened.

He was brief: 'Hello Miss Derbyshire, I'm ringing to tell you your scan is clear – it has not spread to another part of your body.'

'Oh my god Demitrios, I LOVE YOU! THANK YOU SO MUCH, thank you, thank you,' I shouted.

I threw my phone on the bed, raised both my arms in the air, clenched my fists and ran downstairs screaming, 'It hasn't spread! He's just rung me and it hasn't spread!'

That's why I love Demitrios – and it's not just him: I owe my life to compassionate, skilled nurses, doctors, registrars, radiologists, anaesthetists and radiographers from Mumbai, Sydney, Manila, Colombo, Athens, Port of Spain, Scarborough and St Helen's. The sons and daughters of those towns and cities came to London and worked for the NHS. Over 301 days they cared for, treated and looked after me. And I'm grateful for that every single day.

CHRIS EVANS

To risk one's life to save that of another is something I imagine most of us have contemplated. To save the life of someone we know and hold dear, especially our children or children's children. To risk one's life for a stranger, however; I wonder how many of us could even come close to putting our hand up for that?

This is precisely what military personnel sign up for, of course. It's the driving force behind everything they stand for. As it is for members of the fire service, the lifeboat service and the police force, not to mention our dedicated coastal and inland search and rescue units. But it's really not what the hundreds of thousands of frontline NHS staff signed up for. To save lives and comfort and care for the sick and dying, yes. But at the risk of losing their own? No, that was never part of the deal.

And yet look at how selflessly and unconsciously these criminally underpaid, underappreciated, overworked, everyday heroes have cast aside concerns for their own mortality to face whatever new terrors the coronavirus spits at them, on a minute-by-minute, shift-after-shift, day-after-day, night-after-night, twenty-four/seven basis. What these amazing human beings are doing is beyond incredible.

But of course, what's really happening is far darker than that. A lot of our frontline NHS miracle workers are risking their lives not even to save lives but simply to operate various machinery, mostly ventilators and intravenous drips, while nursing patients who are almost certainly not going to make it.

How precisely the virus came about, how much effect the worldwide lockdown is really having and how this pandemic will finally play out is anyone's guess. Frankly, no one really knows anything for certain about the virus itself. If they did, surely they would be screaming it from the highest rooftops. The world is (literally) holding its breath, waiting to see what happens.

However, here's what we do know.

We know that simple actions at the base level of prevention give us the best chance of buying as much time as possible to help those on the front line to help save us. It's not rocket science, but it is science. It's a simple fact of washing our hands, staying home and physically distancing from each other if/when we do venture out. (I don't know about you, but ironically, it seems the more physical space we give ourselves, the closer we are becoming both emotionally and spiritually.)

We also know that most of us are incapable of comprehending what kind of mindset and sense of purpose it takes for our army of NHS brothers and sisters to simply 'keep showing up' and contending with the horror of interminable disease while often unable to do little more than temporarily postpone the merciless inevitability of unavoidable, certain death.

It is, therefore, our unequivocal, unanimous and united duty to let each and every one of our priceless doctors, nurses, carers, consultants and all of their support staff and services know that we will be forever indebted to them and that we love them. That's right, we must show them that we actually love them.

However you see fit, whether it's via a handmade poster in your front window, a flag flying from the roof of your van or a salute every time an ambulance or paramedic passes you in the street, it is your and my duty to LOVE THE NHS.

KT TUNSTALL

Stay.

Stay. Stay at home
While others go,
They leave, they show
That staying is an act of love,
An air-punch pledge to rise above
This tiny thing; invisible,
That's come to change the world.

Stay. Stay at home
For while this sweeping wave will grow,
We grieve, we share, we talk again
And in our words we dream of when
We'll hold each other so completely,
It will change the world.

Stay. Stay at home
Create firm ground for all of those
Who risk their lives to do what's right
To beat this darkness with their light

And tend to those in desperate need
In hope, to change their world.

Stay. Stay at home.

And even when it drives you mad,

Be glad of all the little joys
That wait for us beneath the noise,

And wave this caring army off
To work
To save
To tend
To keep
Our loved ones in this world.

IRVINE WELSH

Don't Even Ask

Well, it was bound to get a bit intense on lockdown at Josh's place. I mean, we'd only been going out for two weeks! It wasn't like we was living together or anything like that. I just ended up there after a mad one and he said: –Might as well just stop here, Cloh, I've got everything we need; grub in the freezer, plenty booze, loads of posh, and Netflix. All this bollocks'll be over in no bleedin time. Old Boris, prime minister, he said it was nothing, shaking hands with all them infected sorts, he was.

Well, it's a nice house with a back garden, better than the tip my flat is right now, truth be told.

I suppose I was still missing Chris. It was my own stupid fault we broke up and I regretted it almost straight away. Chris: so hunky, sweet and loving. Problem was he was absolutely devoted to his old mum. Don't get me wrong, I think that's lovely but sometimes it can get too much, innit. Nah, it ain't right really. But I suppose she is old and she'd had a minor stroke, and he was going to the hospital with her for tests on her high blood pressure. How many bleedin tests do you need, though?

Thing is, I was really premenstrual at the time, and I said something like: –You care more about her than you do about me!

Chris just looked at me and said: –I gotta do my best for her, Cloh. You and I, we got lotsa time ahead of us. She ain't.

Well, maybe I should've been a bit more understanding but I always thought that devious old cow played him like a bleedin fiddle. Anyways, long story short, we had a row. Again, my fault, cos he never so much as raises his voice. And I was the one that ended up walking out.

I met Josh at Caz's party that weekend. I didn't want to go but Ange said it would be better for me than staying indoors moping about Chris. Everybody was sloshed on these cocktails Camp Trevor was mixing and there was loads of gear kicking about. I noticed Josh straight away; he had a bit of a rep as a lad and he was racking up really killer lines of nonsense on the dining table. Call me a dopey slag if you like, *don't you farking dare,* but it impresses me no end when a geezer puts out proper poodle's legs. One thing I can't stand is a tight-arsed bloke. Anyway, when I went to smash one up me hooter, he said: –Get down on it, gel, and gave me this saucy wink.

Later on, when I was dancing with Ange to that Calvin Harris 'Acceptable in the 80s' song that we both *love*, I saw

Josh sitting there with a couple of his mates, drinking Stella and looking over at me. She said: –Oi, Cloh, you know he fancies ya?

Maybe it was the drink and drugs but I gotta say he looked quite tasty in his white shirt with his tan. I'd heard him say earlier that he was back from Tenerife but had 'cheated a bit' and topped up on the sunbed he had at home. I thought: *Sunbed at home? Mmmhmm.* Downside: a fair old gut on him (not like Chris, watched what he ate, worked out, avoided lager) and a bit of a weak chin. But I suppose he could always grow a beard.

So we ended up snogging away and, long story short, started seeing each other. Well, Chris was just taking his mum into hospital all the time, constantly round at hers, dropping off shopping. I suppose I just wanted to make him jealous, to get him to notice me. It wasn't nothing serious, this thing with Josh, just a bit of a giggle.

The next weekend we went to the Printworks for his mate Billy's birthday bash. Pretty messy it was; we was all properly on it. So me n Josh was more or less holed up at his place sweating out the comedown, ignoring all this coronavirus nonsense. I remember him making me laugh when he said: –Coronavirus? What the fuck is that? That's a bleedin nonce virus! Call it the farking Stella virus, then I'd be bothered!

Then this bleedin lockdown kicked in, so I just stayed put. Well, he had the lot in, so we was well set!

But it was too soon really and we started to get on each

other's nerves. Josh has got a great house though, glass coffee table, leopard-skin rugs. –I ain't saying they're real, but I ain't saying they're imitation either, he laughed. I wasn't mad keen on the two giant porcelain apes, standing thumping their chests on either side of the couch, looking onto the 75-inch flatscreen. As I said, he's a bit of lad and he's only got a big collection of these samurai swords mounted on the wall.

We was still a bit shaky so we started doing this cats jigsaw puzzle that he'd got his mum for Christmas but forgot to give her. That sort of set my warning bells off: I mean, you can't really trust a bloke who neglects his old mum, can ya?

Anyway, we soon got bored with that and chopped out a big line and started messing about, as you do. Soon we was properly at it, if you know what I mean. One thing that impressed me about Josh was he never let the ching get to him in the downstairs department. Chris didn't do it a lot, but his went into a floppy little nothing after a toot. Not Josh. But as hard as he stayed, he didn't have much in the way of stamina, even after a big line.

Afterwards, we lay on one of them leopard-skin rugs and I rested my head across his chest. But I was a bit depressed. Not exactly buyer's regret, but it just wasn't feeling like it did with Chris. Anyway, I think Josh caught my mood, cos he got up, his face still all red, and started clowning around again. Racking out more bugle!

Well, I blame the sniff for what happened.

Josh took one of them swords off the wall and started telling me about it. Then he showed me this book about samurai warriors and all the poses they did with the swords. I saw one that reminded me of a yoga position we do in Chell's class. When I told him it was actually called the warrior pose, he got me to hold the sword. I didn't want to at first cos I was off me tits and it was dangerous cos it looked razor sharp. –Go on Chloe, he was all excited: –You do yoga, like you say, it's the warrior pose, shows them things is all connected, you'll be a bleedin natural at it!

I was surprised at how light the sword felt, and how good it felt in my grip. I started swinging it gently and striking the pose, like what's-her-face in that *Kill Bill* film. Josh was having a proper old laugh, encouraging me. It felt great, it was like a wild dance, and I just got really lost in it.

Then it happened.

I was waving it around like one of them Jap farking warriors and then I saw his hand rising up just as my blade swished through the air. Then I felt the very lightest of contact and heard a faint chopping sound.

Suddenly, his face changed.

His eyes bulged right out and his mouth was shaped in a big O.

Then the blood.

On the leopard-skin rugs.

My sister Von being a nurse, I knew exactly what to do: pressure on the wound. Poor Josh was fucked, he was just shaking and saying: –How the fuck did you do that . . .

how did that happen … over and over again as I got him through to the bathroom. Luckily there was bandages and plasters and stuff. As he moaned, I just carried on trying to stem the flow of blood.

Once I stopped it I went through to the living room and saw it, lying on the floorboards, just beside one of them rugs. There wasn't much blood. I half shut my eyes and picked it up off the floor and ran with it into the kitchen, without looking at it. I found a Tupperware box, which I filled with ice, and stuck it in there.

I got Josh out the bathroom and called a taxi. –We're going down to Tommy's; we'll get this stitched back on.

Josh was as white as a sheet when we got into the cab. I told the driver: –St Thomas' Hospital, accident and emergency.

He looked at the hand, wrapped in a bloodstained handkerchief, and went: –Little accident then, what happened?

Poor Josh couldn't speak, looked like he was gonna pass out.

–Don't even ask, I told him.

Well, when we got to the A&E it was proper nuts! People coughing, all them old folks: some of them like death warmed up. Then Josh starts screaming, waving his bloodstained mitt around: –HELP ME! I NEED HELP!

We goes up to this desk, where the receptionist, a geezer, asks him what happened.

Josh waves his bloodied hand in the air. –WHAT DO

YOU THINK! I NEED HELP!

The bloke tries to hand Josh a ticket.

Josh just looks at him, eyes bulging out. –What's this?

–It's your number in the queue. We've a lot of coronavirus cases, and he points over at that coughing mob, some of them not even observing social distance! –Now I just need to look up your records . . .

–You wot? FUCK MY RECORDS!

–Shh, I told him, –You're causing a right bleedin commotion!

–There is a pandemic, the reception fella says.

–THIS IS MORE SERIOUS THAN ANY FARKING PANDEMIC!

Then a nurse wearing a mask and plastic gloves comes up to us. She glances at his hand and the blood on the handkerchief wrapped round it. I waves the Tupperware box at her. –It's in here, and it needs stitched on bleedin sharpish.

The nurse only turns round and says: –I'm afraid he's going to have to wait . . .

–Even if it means losing it? I looks at Josh.

The nurse looks at me, then nods: –We have to observe the new procedures.

Well, I just opened the farking box and showed her what was in it. She proper gasped. Just at that very minute, the blood started to seep through Josh's trousers at his groin. He looked down at it and he started shaking, like almost fitting, and the bloodied handkerchief he'd been holding as a diversion just sorta slipped outta his hand. Of course everyone could see all his fingers was still there!

–Oh . . . the receptionist looked like he was gonna faint.

The nurse shut the box and nodded to the receptionist and we helped Josh through to a treatment room. –You will be able to help me! Won't ya? he was begging her.

She just stayed all calm, pulled the blinds round him. They got to, ain't they, it's the way they've been trained. Our Yvonne told me that. –You really gotta stay calm, she said.

Well, Josh wasn't staying bleedin calm: tell you that for nothing. Cos the nurse ain't talking, he's asking to me: –Am I gonna lose it, Cloh? Am I gonna lose it?

So I had to stay composed for us both. –Course not, they'll sort it, it's the NHS, I told him. –Bleedin miracle workers, they are!

It was only about half of it.

This doctor came along and they said I had to go. So I left Josh there, sitting on the bed, tears rolling down his cheeks. What got me was when the nurse asked: –Who are you to him?

I couldn't even think of what to answer.

So I took a seat back out in the waiting room and tried to ignore all that coughing in my ears from them bleedin spreaders. That A&E must be the most infectious place in London!

Then one of them old geezers was looking round at me.

–Wot? I said, and he turned away. Proper creepy old

cunt, he looked like that old perve in *Game of Thrones*.

Funny thing was, that fat, useless ponce of a prime minister was upstairs in intensive care, from shaking hands with all them cunts wot had the Billy Ray. What do they teach them at those posh schools besides noncing? Anyroads, I wasn't sticking around here to breathe this air. I thought I'd best scarper to the cafe for a cup of tea.

When I got there, the place was shut: no social gatherings, only a long queue for a poxy vending machine. Might have made an exception for a farking hospital is all I can say.

But then I got a proper old surprise when I saw who was there, first in line. It was Chris! He nodded at me and got us a second cup of tea. I was absolutely made up to see him, but he just looked so sad. –She's gone Chloe, his eyes were blinking and tearing up. –Me poor old mum. Virus got her, he said, all sniffy. –Didn't stand a chance.

I really, really wanted to hold him but I couldn't, not with all this social distancing lark. But I let him know that he shouldn't be alone, not at a time like this. Told him I was going back with him to his place.

We got outside and there was cabs waiting so we climbed into one. Just as well, cos it was starting to rain. On my life, I said there and then, in the back of that bleedin taxi, I'm so sorry I was such a bitch the other week. Told him what a great woman his old mum was, proper salt of the earth.

Then I thought about poor old Josh, but at least he was in the best place. Hopefully they'd be able to do something for him. They really were bloody heroes. He would be OK to find his own way back home: there was loads of cabs. The way I was thinking was that Chris was the one what needed looking after: devoted to his old mum he was. You can imagine; predictably, he was in bits.

–What you doing here anyway, Cloh? he asked, his big, dark, dreamy eyes glistening.

I shook my head and settled back in the seat. –Don't even ask.

THE HAIRY BIKERS: DAVE MYERS

Dear NHS, an Apology

Dear NHS,

I'm afraid to say I was a bit of a twat and a drain on resources when I was a child. For this I must apologise. I was born in 1957 and it really started before then. My mother went to see Dr Morrison, a serious Scottish GP (I incidentally thought he was Dr Finlay), who was working in my hometown of Barrow-in-Furness. She was diagnosed with an ovarian cyst. She duly presented herself at North Lonsdale Hospital, only to be told, 'Well Mrs Myers, it's the biggest bloody ovarian cyst I have ever seen. You're pregnant – it's not a cyst!'

I was to be my mother's first child at 41 and she was over the moon; I, at this point, was none the wiser as I was just a chuffing cyst. She had high blood pressure and was confined to hospital for the last ten weeks of her pregnancy; I feel no guilt for this, as I had nothing to do with it – it was my dad's fault.

Nevertheless, thanks to the NHS, I was duly delivered by Caesarean section some nine months later. As my mother always said, Caesarean babies are more intelligent. My dad, however, got a bollocking when my mother got home as she realised that, during the ten weeks she had been in hospital, he hadn't changed the sheets. But for my birth, I thank you.

The first time I troubled you on my own account was when I was two years old. I can remember being food-obsessed even then. My mam's fishcakes were the best. They were made from hake, with lots of buttery mash with white pepper and were covered in those orange breadcrumbs that you can see from space. Then they were smothered in an equally fluorescing cheese sauce ... ooh, I can taste them now. As Mam cooked, I sat cross-legged on the floor, playing with my wind-up clockwork monkey that beat a drum. I sucked the key in anticipation, then, with a massive suck and a drool, the wing-nut-shaped key slid down my throat or, more accurately, got stuck right across, I believe, my windpipe.

'Mam, I've swallered the qwey.'

'What have you swallowed?'

'A qwey.'

Looking down, she saw my much-chewed box of Crayola crayons.

'Ah you've swallowed a crayon; don't worry, it'll melt and you'll pass it.'

'Nay, the qwey!' I remonstrated, pointing at the empty keyhole in the toy monkey's backside.

She went hysterical and realised that one bad move could see me choke. 'Put that bloody monkey down and come with me.'

She shepherded me out of the back yard and down to Mr Gibson, the local chemist. Mr Gibson, a wise old soul, said, 'Oh shit, get him in my car.'

A car! Wow, this was so worth it. I had never been in a car before – this was 1960 and we were a motorbike and sidecar family. The car was amazing and it didn't even matter that I had to listen to my mother howling. Though I knew that the monkey was no good without the key, which was my big worry.

I've got vague memories of the hospital. I was kept in overnight, given a general anaesthetic and they removed the key with forceps. I was in a ward with other kids, lots of whom had had their tonsils out. Like them, I got loads of free ice cream to soothe my throat. On the locker next to me was the key and there was a little X-ray picture of it stuck in my throat. I was overjoyed – what an adventure! I returned home, this time sadly on the bus. When I got home I headed straight for the toy box, only to find that my drum virtuoso monkey had been put in the bin for my own good. So thumbs down for the parents, but a big thank you to the NHS.

The NHS was there throughout my childhood. There was the time I tried to build an aeroplane out of a wooden crate and old scrap. I was seven at the time. The backyard wall was about two metres high. I hauled my plane onto the wall and I was sure that as I pulled back on the joystick – well, an old toilet brush – I would soar over the rooftops. I rocked backwards and forwards until I plunged to the concrete yard. My dad picked me out of the wreckage and put me on the back of his motorbike and rode again to North Lonsdale Hospital. I was put back together with two splinted and plastered thumbs. I worried all the way home how I was going to wipe my bum. Thank you, NHS.

The next year, I was back with a broken nose when one of the kids in the backstreet put a fishing cane through the front spokes of my tricycle as I rode along. I pitchpoled over the handlebars and broke my hooter. This time there was a lot of blood. Back to North Lonsdale. I came home taped up, complete with two black eyes. I don't think it ruined my good looks though. My mam said I ended up with a Roman nose . . . roamin' all over me face! Thank you again, NHS.

There were two other occasions for which I have to thank the NHS for their help. One was when I was about ten years old. Ecstatic at completing my homework, I jumped up with glee, landed and impaled my foot on a pencil. I was skewered. My dad, with his northern grit, simply pulled out the pencil. The lead broke off and was left embedded deep in my ankle. Some weeks passed and my foot started to abscess, so I was admitted for an operation to remove the lead. This time I got a really big bandage and stitches. As I danced on *Strictly Come Dancing*, I thanked the NHS for saving my foot.

The second time, when I was twelve years old, I was properly rushed to hospital and my life was saved. I'd had really bad stomach ache. Oh, how I whinged, moaned and complained. My mother, never really Florence Nightingale, put my condition down to the fact that I had just eaten loads of mushy peas and cheap sausages.

'Jump up and down, son, it'll break the wind. Have a big fart and you'll feel better.'

Being obedient, I bounced up and down like Tigger, which wasn't easy, given my recently repaired foot. I jumped and then the pain suddenly got worse. I had bounced until my appendix had burst. Dr Morrison appeared within the hour. He was serious and phoned for an ambulance. As I was stretchered out of the house, the street came to watch. I had quite a crowd; I loved that, and I haven't changed really. During this bells and blue-lit journey, the great ambulance man kept his eye on me. I was worried people could see me as I was going along but he said the glass was magic.

'You can see out, but from the outside it just looks black.'

I thought this was brilliant. I remember being prepped for the operation – at this time I didn't know what peritonitis was, but I heard the word mentioned. I had the pre-med injection and loved it. I thought that it must feel like this being Batman. This certainly set me up for a good time during the seventies and art school!

The next morning, I woke up sore but feeling very important and the wonderful nurses treated me like a prince. I thought that they must know that things were hard at home, as during this time my mam was ill with multiple sclerosis. I thought that they were being extra nice to me but they weren't – they were angels like this to all the kids.

I was fascinated by a lad with a catheter and a bag. I asked him what was going on.

'It's just me piss bag. I can empty it myself, you know.'

I wonder what happened to him?

After a day or so, I got transferred to a convalescent hospital in a village called Aldingham. This place was like a stately home. To a child from a red-brick two-up, two-down, whose horizons went as far as the backstreet, this was heaven. The food was great and every day I could go onto the beach. The downside was that twice a day I had to have antibiotic injections of penicillin and streptomycin. These names are engraved in my memory over fifty years later. Back then, these were big injections and the syringe seemed huge, and yes, it hurt. One nurse was more at home playing darts down the pub than dealing sensitively with a vulnerable youngster; I dreaded her doing the jabs. My arse ended up like a battered mango.

This was when I learned not to call the surgeon 'doctor', as they were called mister. Two of the grown-ups were talking; they were discussing the various merits of the two surgeons who worked at North Lonsdale Hospital. They said Mr Daniels preferred butterfly stitches or clips, which were easy to remove, and when he did do stitches they were small and tidy. The other, Mr Easton, they said doubled up with his stitches, which were massive, and he didn't believe in clips. I didn't know which surgeon had dealt with me. So, with trepidation, I looked down at the clipboard hooked at the foot of the bed. I was one of Mr Easton's patients. I pulled down my pyjamas and looked at the huge plaster. I wondered with horror what lay beneath.

I counted down the days – good food, horrible injections, nice people – and then the time came to remove the stitches. I panicked – how on earth do they get this industrial-sized plaster off my belly? I soon found out, as my hated nemesis, the dart-playing nurse, gave me a snarly smile and ripped it off. I did a little wee.

I looked down. My belly looked like Frankenstein's head. I was a bit of a plump child but surely my chubby little belly didn't need to be sewn together like a ripped rugby ball. It hurt getting the stitches out. Thank you, the NHS, for saving my life.

To conclude, I really do mean a big thank you to the NHS. Throughout my childhood you were always there for me, you listened and never let me down. I grew up thinking it was my right to be looked after and to be healthy. How lucky was I. Thanks to you I am still here.

Let's bring back the times when the NHS was financed properly, appointments were easy to get and waiting times were short. I hope we learn from this grim situation and the huge dependency we have experienced with coronavirus that we must allow the NHS to grow, and to continue to be the envy of the world it always has been.

THE HAIRY BIKERS: SI KING

As someone with a propensity to be accident-prone, and who has travelled around the world on a motorbike, I've had the privilege of seeing many of the world's healthcare systems. Now, some of those healthcare systems are good and some are bad, but not one of them, in my experience, compares to the NHS. I am so very, very proud of the egalitarian system that we have in the UK and the attitudes, characters, skills, intellect and expertise that are applied to care in our country.

As some of you may know, I had a brain haemorrhage six years ago.

I was cared for by the ITU team at the Royal Victoria Infirmary in Newcastle. They were THE MOST kind, caring, professional people I've ever met. Newcastle being the small city it is, I've seen some of the care team that looked after me out and about in town. We always stop and say hello. It's a salient moment when you're that poorly and vulnerable and you're totally reliant on the care and kindness of strangers. They are no longer strangers to me.

It just remains for me to say a huge thank you to everyone who works in the NHS, from consultants to cleaners. Thank you, thank you, thank you. You are a massive part of the character and identity of the United Kingdom and we are all rightly proud of what you do.

We all love you!

P.S. RVI ITU, there's a Belgian chocolate torte coming your way – I promised you one.

ROB BRYDON

When my oldest son was just a boy, around eight or nine years old, we were walking together by the river when he fell over and cracked his head open. Blood began to spurt and I knew instantly that the situation required attention from people with greater medical knowledge than mine. Kingston Hospital was the nearest and so I scooped him up in my arms and ran with him to the car. With my son safely in the passenger seat, and holding a clump of tissues to his forehead, we sped off.

Once at the hospital he was treated with great care and compassion and soon my patched-up boy and I were heading home.

Whenever I remember the incident I'm always filled with guilt.

Not because he had the accident while under my care – it could have happened anywhere – but because while I was running to the car with my bleeding son limp in my arms, my primary thought, one that I couldn't get out of my head, was how much I must resemble Dustin Hoffman in *Kramer vs Kramer* when he runs through the streets of New York with his fictional son in his arms. Even while we waited in A&E I was picturing myself running through Manhattan. I might even have tried the voice, quietly, under my breath.

Oh dear.

DANNY WALLACE

We are tired but we are thrilled.

We have a daughter. She's tiny, she's wondrous, she's beautiful. I am now a father of two, yet like a lot of men I'm not quite sure how this happened. I know I'm a dad, but I still feel like a son. Mainly because I still like video games and films with Jason Statham in them.

My folks come to town, and my wife's mum too, and we decide to spoil my son for having a stinky little sister. So we take the train into the city – me, the two grandmas and him.

We find our way to a children's party and then we eat sandwiches at the Waldorf, because that's the kind of thing grandmas like to do, and it's just as we're on our way home that the worst happens.

I turn, carrying my son, to see my mum catch her foot on a paving stone and begin to fall forward.

Like in a dream, the rest is in slow motion, as I work out she's not going to be able to right herself, and I'm too far away to stop it happening, and she falls further forward, and she can't get her arm out in time, and as my heart starts to pound and my body is shocked into action, her head slams hard on a grey city slab.

I shout and start my dash towards her, still clutching my son, and as I get her up the blood starts to pump from her forehead – bright, fresh, red blood, almost neon somehow – and it gushes and runs and covers her eyes.

We are outside a pub and I yell to some men to fetch water and serviettes to press on the wound, and I feel for my phone – because this is an ambulance situation, right? I ask her a question but she's too stunned to answer, so we start to lay her down.

A group of women pound towards us, all comfort and concern, and it is then that I realise I am still holding my boy, who is quiet and timid and staring.

I dial for an ambulance as these wonderful, kind women hold my mother's hand and I move my son away but make sure Mum knows I'm there, as she lies on this London pavement, surrounded by blood and chewing gum, smokers standing around outside the pub, concerned but still dragging on their fags.

'How old is the lady?' asks the woman on the end of the line.

She's not old, I think. *She's my mum*.

But, 'She's seventy,' I say.

My God, when did that happen? She's *seventy*.

'Help's on its way,' says the voice and I kneel by my son.

He shouldn't be seeing things like this. I can see he's scared. This doesn't happen to grown-ups. Grown-ups aren't the ones who need help.

'I promise you,' I say, looking deep into his eyes, 'everything is going to be OK.'

Finally the bravado crumbles and this little boy's voice cracks as he says 'OK', and he melts into me, trembling, his head pressed deep into my neck. He clings to me, squeezing me tight, and we move to his grandma so he can hold her hand.

The landlord of the pub is out now and one of the waitresses asks if she can bring us anything. Someone produces an umbrella to shield my mum from the sun and a passing nurse kneels down to check on her. My mum doesn't want to be a bother. She asks the women who have stayed with us where they're from. Turns out they live on the same street as my cousin in Carlisle. We all try to laugh.

Finally – the ambulance. The staff are calm and kind. Inside the van, they make my boy a blow-up elephant out of a grey surgical glove and take his blood pressure just for fun. He starts to smile again. They're at the end of a twelve-hour shift on a Saturday in London and still they do this.

At the hospital, I could cry when I see how kind they are to my battered, bleeding, broken mum – this woman who has only ever been kind to others. How does our NHS work? How can these people be so warm, so reassuring, so good at their jobs on shifts that long and wages that low? Why don't they sigh and treat us with contempt, the way people in the bakers or the department store or the garage do for something as slight as not having the right change?

And how have these people, this team – from Thailand and Poland and Ireland and the UK – made this promise to the world, that when absolute strangers are in trouble, it will be them who'll be there to pick up the pieces and tie off the stitches and reassure?

My mum broke bones and earned scars and completely *ruined* a pavement that day. But I hope in time my son will forget that and remember instead that what she'd always taught me is true; that people are good. In a crisis, their instinct is to help.

So Carlisle girls: thank you. Passing nurse: thank you. Guys at the Leicester Arms: thank you.

And to the tired, happy staff of the London Ambulance Service and the NHS: *thank you* and *how do you do it?*

It's a horrible moment, the moment you have to care for a parent. It does not feel natural and yet it is second nature. I still don't quite know how I feel. I stood there with my mum and my son and I was both man and boy; I was both father and son.

But I do know this: I don't think there's ever been a day I haven't told my mum I love her.

And now more than ever I know there won't ever be a day I don't.

DAISY MAY COOPER

The birth of my daughter, Pip, in 2018 was hell. I ended up giving birth by emergency C-section under general anaesthetic and the doctors and midwives saved my life. I was eternally grateful for this, but not as grateful as I was when a lovely nurse helped me to have my first poo after surgery.

Constipation is a pain in the arse (literally), but I have never known anything like this. I was recovering from the surgery in hospital and it was now the fifth day of no poo. I lay in the bed, colon full of compacted shite that actually felt like it was fermenting inside me. Awful poisonous farts would escape and smell like the toxic gas from a mummified corpse; they freed up a little space but still no relief.

My family and friends came to visit and to coo over the new baby, but I didn't care – I smiled and nodded when they said how beautiful she was, but it was all a mask. All I could think about was my arsehole giving birth to the two-metre-long anaconda snake of a shit I had inside me. Now that would be beautiful.

I was tormented; the sixth day passed, then the seventh. It was agony. I forgot what it was to be human. I had taken such a simple bodily function for granted – what I would give to have a dump. It was so bad that I even considered phoning up the Samaritans and telling them of my plight. I would have given my life savings for the cure. Rocking back and forth, standing up, bending right over so I could touch my toes – nothing helped.

I was too scared to eat anything – all I could think of was the seven-day-old Greggs tuna baguette still up in there. Childbirth had been hell, but this was worse – this was purgatory.

On the eighth day, a nurse entered the room to do a routine check-up and give me my meds. As she spoke to me I could feel a hot fat lump in my throat and the tears began to roll.

'Are you OK?' she asked.

'I haven't had a poo for eight days,' I bawled.

'Oh my goodness, you should have said so, I'll fetch you some liquid laxatives.'

She returned with the laxatives and, after half an hour, I made my way to the toilet. It was the greatest shit I've ever done in my life. That angel in blue had saved me, and I cried on the toilet in gratitude.

After I came out of the bathroom, I lay in my bed and enjoyed cuddling my daughter for the first time, gazing into her eyes. My god, she really was beautiful.

My husband entered the hospital room with a Costa coffee and a *Take a Break* magazine that he wafted in front of his nose as he retched.

'Jesus Christ, you finally went then,' he snorted.

DOLLY ALDERTON

I've been rushed to hospital once in my life. When I was six years old after an unfortunate incident on a seesaw. I was with my nanny while my mum was at work, thrill-seeking in Highbury Fields playground, pushing the health and safety limits of the apparatus as far as I could in the pursuit of a good time (old habits die hard). I asked my nanny to plonk herself down harder on the seat of the seesaw to help me fly off my seat, just for a nanosecond. With trepidation, she did. 'More!' I demanded. 'MORE!' She slammed her arse down and the handle of the seesaw went into my chin, splitting it open and knocking me straight off the seat and onto the ground. It is the first and only 'cut to black' moment of my life that I can recall (other than the time I tried absinthe). I felt the force of the cold metal in my face, then I saw the terrified, open-mouthed expression of a woman who had accidentally injured a child who was in her care, then I saw darkness.

The next thing I remember is two men who looked a bit

like the Chuckle Brothers, one of them bandaging up my chin as I lay on the ground, the other scooping me up in his arms and walking me to the ambulance. I remember the man in the uniform doing silly voices and impersonations of my favourite TV characters all the way to the hospital. I remember smiling faces in the ward. Some stitches that felt like little pinpricks. My mum taking me home for fish fingers and oven chips. A good war wound and story for everyone at school. And a pinky-silver wobbly line of a scar that remains on my chin to this day.

Except that's not what happened at all. I have since been told that it was an absolute horror show, my denim-blue dress soaking with blood as I lay crying on the asphalt in pain. I couldn't speak, both out of shock and because the blow to my chin meant I was unable to move my face. The two men in the ambulance looked, apparently, nothing like the Chuckle Brothers. 'I could hear your screams in my office and I rang home immediately because I knew something had happened,' my mum says now – an example of her skill of telepathy that she sometimes likes to boast about (she also used this apparent 'telepathy' as a threat when I was a teenager, telling me she'd know if I'd lied to her about where I was going out).

And yet, I remember almost none of that. The kindness of the NHS workers, their commitment to keeping a terrified little girl calm and distracted as well as safe and cared for, worked. It eradicated nearly all memories of trauma. It is a minuscule moment in the grand scheme of what NHS staff have to see and do for their jobs – I am incredibly lucky that I haven't needed to go to hospital for anything more serious. But I will always remember the staff who looked after me that day and feel utter assurance that no matter how big or small the reason that someone might need the NHS, we are always in safe and capable hands.

MARK WATSON

Dear Vlad,

It's me, Mark. Mark Watson? Yes, sort of a minor comedian and author. *Taskmaster*, *QI*, that sort of thing. *Rugby's Funniest Moments*. What? Are you sure? I thought *everyone* watched that.

Nothing? Oh. Well, I'm pretty confident you'll remember me when I tell you about the day our paths crossed, in Archway, north London. In the theatre for emergency surgery, or whatever frightening name flashed past my eyes on the way in. It was a sunny afternoon, late May 2014, although the weather was something of a moot point as we had been in the hospital for twenty-one hours when I met you. We were waiting for our second and final child, our daughter. I suppose 'we were waiting' doesn't entirely do justice to the division of labour, no pun intended. My wife was going through the most gruelling physical process available to humanity; I was mostly fetching bottles of water, pacing up and down and – my main skill in this or any crisis – hoping for the best.

Due to this endless, exhausting stalemate between the baby and her mother, it had been decided the delivery was going to need an emergency caesarean – as had happened with our first kid. Once more, the corridor full of scrubs, the lonely room high up in a hospital wing, the frightening arsenal of metal equipment. A sheet hung up to separate our heads from the butchery that was about to take place. I tried to digest the idea, just as incomprehensible as it had been the other time, that this daughter already existed, had existed for months of course, and in half an hour would enter our lives forever – but for now was in this bewildering waiting room between nothingness and planet Earth. Bewildering for us to think about and, to be fair, probably a little bit strange for Rose Watson, aged minus twenty-five minutes, as well.

And there you were, Vlad. 'I am the consultant anaesthetist,' you said, 'and you are the father?' 'Yes,' I agreed, even though I was so addled that I probably would have said yes if you'd claimed *I* was the anaesthetist and put me in charge of the epidural. Sensibly, though, you did not do that. Instead, you began the process of injecting my wife at the base of her spine.

While this was going on, you chatted to me about the fact you were from Ukraine. I had been to Ukraine a couple of years before, for a football tournament. With adrenalin surging through me, with my brain grasping desperately at distraction, I gabbled at you on this subject like an over-eager *Just a Minute* contestant, supplying what was probably a very full breakdown of the game (a 1–1 draw between England and France), the goals, the layout of the stadium and the night my brother and I spent on the floor of Kiev airport before flying home. Yes, surely you remember me now. Surely you remember thinking, 'How fortunate I am, that the husband of one of the day's final patients is in a position to give me such a precise summary of a group fixture in Euro 2012. Yes, lucky old Vlad!'

'Everything is going good,' you said, 'everything is fine.' You said this several times as the rest of the team began work; your tone of voice was as casual as if you were welcoming us on board the service from Paddington to Cardiff Central. This was even more valuable than you probably realised. I, after all, was trying to give exactly the same reassurance to the mother, but it didn't mean a huge amount coming from me, any more than in turbulence it is worth me saying, 'Oh, I'm sure we aren't going to crash or anything.' I had absolutely no idea what was going on. But you did, and you made me feel bolder than I really was.

By now – with the epidural fully kicking in – it wasn't clear to me whether you still had actual clinical tasks in the operation, but what I did know was that you had now become a full-time helpline to me and, by extension, to the mum. 'You like music?' you asked and, without warning you switched on a small, old-fashioned transistor radio. The birth of the first baby had taken place with only the quiet chat of the surgeons and the clinking of unspecified scary tools behind it and I think I'd assumed that was how things always went. The birth of a child didn't seem, in my head, like something you deployed massive beats for. But it couldn't have been more welcome. You yourself seemed very into it. Your face relaxed into a grin. You looked moments from shouting 'EVERYBODY GET ON THE FLOOR!!'

'You like the tune?' you asked – remember, Vlad? – and, yes, I said, I very, very much did.

The song was 'Once In A Lifetime' by Talking Heads. Always a favourite of mine, but never quite as perfect a soundscape to anything as it had suddenly become. The celebrated opening bars with David Byrne's spoken-sung sermon, concluding in '. . . and you may ask yourself: Well! How did I get here?' What an amazing note, I thought, to enter the world on. But *was* she coming?

'Yes, yes, everything is going good,' you said once again. 'Everything is absolutely normal.' I clung onto that word, 'normal', like the life-raft it was in these dizzying circumstances. With whatever small recess of my brain was still processing actual thoughts rather than just blasting alarm bells, I thought: 'How could this *ever* be normal? How can it be that your work day consists of nothing but people like me – maybe more composed than me, as almost everyone in the world routinely is, but still people on the precipice of one of the defining moments of their lives? How can you possibly handle six "defining moments" in a shift without the stakes overwhelming you?' I know it wasn't your child, Vlad. (At least, I'm pretty confident it wasn't.) But to be so calm that you could transmit calm to two scared people; three, if you count the emerging baby, who was having a stressful Monday of it herself. It's quite something to be able to do that, Vlad.

If it had been a Netflix show, 'Once In A Lifetime' would still have been playing as Rose spent her first few seconds in London and on Earth. But not even your team could complete a C-section in less than the playing time of a single – unless we'd struck lucky with 'Stairway to Heaven', I guess. The song that followed it was 'There She Goes' by the La's. Again, not bad: a woozy, iconic anthem, seemingly in praise of a girl – but unfortunately, I don't know whether you're familiar with it, Vlad, it's widely known to be about heroin. 'You like the tune?' you asked again, but I didn't get a chance to answer, because suddenly one of the midwives had my new child in her arms.

The few seconds of suspense, and silence, when the baby first appears are like nothing it's in my power to describe. You still know anything could be wrong; disasters happen, unforeseen ones – you will have seen plenty, Vlad. Then she opened her mouth and wailed. 'It's a lovely, healthy big girl,' said someone. I was crying. My wife was crying. Rose was loudly making her initial impressions of the planet known. 'There you are,' you said, Vlad. 'Everything is good, right?' And it was. Like you'd said all along.

So, all of this ring a bell? No, I guess not. Like I've said, to you this wasn't even your first wonder of the afternoon. You saw more babies born that *day* than I have seen in my life. And the better things went in each case, the less you will remember about them. For all I know, even the music wasn't some unique choice of providence; maybe you have a playlist that you always pop on, to make every dad feel like there's something special about him and his experience. If so, fair play, you judged it to perfection. And I hope the Super Furry Animals are on there somewhere a bit later.

So, in fact, you won't have had a single thought about me since 26 May 2014, or my then wife or our now six-year-old, delicious daughter. But I've thought about *you* plenty. Every time Talking Heads come on, every time I drive past a hospital, every time someone talks about 'going into theatre'. I don't know if you still live and work in this country. For all I know, you might be a painter and decorator now. Or, I guess, a DJ. But that's the point: you don't even need or want me to know anything about you. You represent the thousands of people in the NHS whose names we barely learn, to whom we make no special gesture other than a muttered thank you on the way out of the room and who, nonetheless, change our lives as a matter of course.

All the same, I didn't change your name, or invent any of this. You *were* called Vlad. You were working in a north London hospital on that day in May. And maybe, just maybe, you'll see this, or someone will somehow work out it was you and tell you about it. And this will maybe serve as some sort of thanks.

Mark Watson (from *Taskmaster* and stuff)

KATARINA JOHNSON-THOMPSON

When I was in year three, the NHS visited my school. There was, and still is, to my knowledge, vision screening for all children between the ages of four and five years. An optician comes into the school with the purpose of identifying any child with reduced vision.

I was always very competitive and I remember the eye test almost like a game. I came out convinced I had beaten it and I had no reason at that age to believe that my vision was anything other than normal. When I met my friends after, we all gathered around, comparing letters we had seen. 'I got W – R – O – X for the last line and the woman said "perfect". What did you see?' I answered 'Same,' knowing that I'd had a nightmare and things that I had really struggled with my friends could see with ease. Sure enough, the next week my mum got the letter confirming I needed glasses and to come into my local opticians for a proper test.

Before you know you have reduced vision, you have

no concept of what you should and shouldn't be able to see. You don't know the difference between having to sit painfully close to the TV so you can see what's going on, or not knowing if you should stick your hand out for the bus because you can't see which number it is until it's almost too late. Or why you shouldn't play hockey because you won't be able to see the yellow ball against the green Astroturf and you might whack a teammate in the face by accident because you have no concept of space.

You don't know any better when you're young, but by the NHS going into schools and funding these schemes, it gave me confidence. I found out that things I had struggled with were because of my vision and not because I was bad at them. Getting glasses meant I could take part in sports, take notes from the classrooms better and interact more socially. I can't say how I would have progressed without this intervention but I know that after it I was a much happier and more self-assured child. So I'd like to thank the NHS. With their help, I was able to apply for free glasses and have them throughout my education and formative years, something that would've taken me years to pick up on and could have been a struggle for my family to afford.

MARTIN FREEMAN

My contact with the NHS started early. As a young child, I was diagnosed with asthma, so trips to the doctor's surgery and hospital were not infrequent, but they were scary. More for my folks than me, I think.

Then there was the dodgy hip. Great Ormond Street and the London Children's Hospital. Then the Rowley Bristow Hospital in Surrey. Good times.

Then the appendectomy – Kingston Hospital – only a few months after another op at the same place which was too embarrassing to inform my schoolmates about, so it went under the heading 'bad hip again'.

Also the close family members who have spent their last days in the care of the NHS.

And the emergencies with our kids when we can't exhale as we worry about what's coming next. Nothing at all unique in that. We've all been there, in one version or other.

Which is what makes the NHS the thing that most of us

would fight tooth and nail to save. The NHS is what makes us truly proud, along with a few other things, to be British.

'Cradle to grave' was the idea, wasn't it, and my God, ain't that the truth.

GREG WISE

Safety Third

I have a large, blood-red lettered sign hanging in my workshop at our cottage. It reads 'SAFETY THIRD'. These two words pretty much sum me up as a person. Throwing myself headlong into most things with a great amount of energy, but with less self-care than really ought to be present, leads to regular blood loss. Which, any rational person would have thought, would then involve a health service undertaking the various bits of stitching and resetting of bones required by my lack of personal care. However, combined with my SAFETY THIRD policy, I am also a Northern Man – unwilling to seek professional medical help. In my view, most things can be dealt with by a pile of loo roll taped around whichever limb is leaking, or for bones a good brisk walk will sort it out.

Maybe I have a genetic predisposition to the avoidance of hospitals. My dad was an architect and one of the projects he undertook was a new wing on an infirmary in Newcastle. It was quite a difficult gig for him as he had a tendency to drop into a dead faint whenever he walked into a hospital. I think it had to do with the smell. Or the pervading sense of mortality. It was not a good look for the professor to keel over on a site visit.

The moment I knew that my wife was 'the One Able to Live with Me' came as a result of a wound: I wandered up the garden, shortly after we had started living together, with a very large hole in my leg – a blood-dripping, bone-showing sort of hole – having just fallen out of an elder tree, where I'd been collecting flowers to make elderflower cordial. She calmly went to the medicine cabinet to see what was on offer, unfazed that her thirty-year-old feller was still climbing – and still falling out of – trees. It was suggested that a trip to A&E would be a rather good idea, but Northern Man looked at the total lack of skin around the hole and came to the conclusion that, as there was nothing left to stitch together, it would be a waste of everybody's time. A bandage wrapped around it would suffice. A few years later, during a hot shorts-wearing summer, a doctor mate looked at the impressive collection of scars on my shin and asked how long I had been in hospital for my pre-tibial laceration. Allegedly I really ought to have got septicaemia; could have had my foot/ leg amputated. Could have died.

I sliced through to the bone of a finger while my daughter was reading *The Beano*, waiting for Dad's Famous Schnitzel. She was quite young and my wife was away working, so, once again, Northern Man, hand held high above his head, blood running down his arm, wandered up the street to the local pharmacy. I was told to go and get it stitched at the hospital. I didn't go, but bought a large collection of bandages. I have had no feeling in the end of my finger since then.

I also pushed a chisel right through another finger. Same story.

There's a bit of a recurring theme here: Northern Man doesn't like fuss. Doesn't want to burden our stretched health service needlessly. And maybe, somewhere deep down, there's a slight sense of shame that I'm just so bloody accident-prone. That by turning up to a hospital with an idiotic self-inflicted injury, I'd be viewed by the stressed medics rather like a dog whose head is stuck in the park railings: an object of pathos, humour, but mainly just a right royal pain to have to sort out. And Northern Man has his pride.

All this changed totally when my sister got ill. Northern Man was humbled and now prostrates himself at the feet of the most extraordinary people he has ever had the privilege to meet.

I smashed the knuckle on my left-hand index finger in a cricket accident. Just as I was about to cycle to my dentist's surgery to get it x-rayed (avoid hospitals at all costs, and my dentist's has a perfectly adequate x-ray machine, doesn't it?) my sister Clare called. She told me she had been diagnosed with bone cancer.

For the next number of months our lives were pretty

much spent in hospitals – notably the Fracture Unit at University College Hospital, London, where Clare was subjected to a large number of x-rays, CT scans, MRIs and surgical procedures. At the start of every consultation, I asked whether the attending physician could take a look at my finger. No chance. Not without a 'blue slip', only available from the referring doctor or via a four-hour wait at A&E. Could I not come in with my sister to this next scan and place my finger next to her fractured humerus? No chance.

Finally, eight months after my accident, I succeeded in wearing one of them down and they filled me out a blue slip and sent me down to the Cancer Centre Imaging Department.

'What type of cancer do you have?'

'I don't have cancer.'

'What are you doing here, then?'

'Um, cricket accident, but I *do* have a blue slip . . .'

I had an operation the next month. A long, jagged cut on the palm-side of my finger and a pin put into the joint to keep it stable. It went septic. A week later, lying on the couch in agony in the doctor's room, him trying to dig out the pin that had been swallowed up by the swollen, septic flesh, I had to ask the nurse to please stop grimacing and I really didn't need to know that this had put her off her lunch.

My finger looked like a fat German sausage. I couldn't bend it more than a few degrees. I was told to be very careful with it.

Instead I built a shed and painted it. At almost the last brush-stroke, balanced at the top of a wobbly stepladder, I fell, hit the flagstones and then carried on falling down a flight of brick steps until brought to rest by crashing against the garden gate. My finger was fine. But I'd fractured the thumb next to it.

My sister bandaged and arm in sling; her carer brother with two digits in splints. Often, when checking in at the hospital, the receptionist wouldn't know which of us was the patient.

There then followed three months of being my sister's live-in, 24/7 carer. And for the first time, I fully understood what an incredible gift we have been given in our country – the NHS. The Gift That Keeps Giving. From the receptionists to the nurses, oncologists, radiographers, therapists – everyone we came into contact with met us with kindness, compassion, humour, expertise and, above all, a profound sense of humanity.

Our local hospice was fundamental in helping my sister stay, and eventually die, in her own home. All the kit they arranged; sitting down with Clare and gracefully and kindly explaining where she was in the progress of her disease, and in supporting me. I would turn up at the hospice every Friday over the last months – a sweaty mess, having just cycled up the hills – to collect Clare's medication for the next week. I would be pulled into one of the doctors' consulting rooms, sat down and asked how I was faring. I was terrible Northern Man to begin with – embarrassed that I was the one being focused on when they should be directing all their attention to my sister – but found myself profoundly grateful to be asked. Because, as we see only too clearly now in this pandemic, we need to attend to the caregivers as much as those they care for. The wheels fall off when the caregivers get sick.

Over the months of caring for my sister, I lived pretty much in isolation. She would be asleep for most of the day, so I was alone, unable to leave, with a constant bubbling anxiety. But this time allowed thoughts to form, gestate and be expressed – thoughts that I would not normally have had the time to explore in the mad rush of normal life. And we, as a nation, as a World Family, now find ourselves in a similar isolation – unable to leave our homes and with a bubbling anxiety. And at an odd moment, when we have to be as self-reliant as possible, but also when some of us will require more help than we've ever needed. From our health service. From the people I have always tried my hardest to avoid.

It took the dying of my sister to fully appreciate the wonder of our NHS. In this time of pandemic, we are, as a nation, joining together to applaud these incredible people. And I (for the moment at least) have altered my habits and now try to operate on a SAFETY FIRST basis – doing my bit for the NHS by not turning up with bits of me hanging off.

But I think it's OK for us all in this time of lockdown to be a little bit Northern Man. Stay home. Stay safe. Wrap a pile of loo roll around whichever limb is leaking, or for bones, a good brisk walk will sort it out. So, as I type now with my two sensationless fingers, one pointing in a very strange direction, I ask all of you: please protect our incredible NHS. And let us never forget how we feel about them, now, when we need them. Let our support grow and grow. Treasure, nurture and applaud forever.

KATE MOSSE

'Keep Smiling Through ...'

At a quarter to eight on Thursday 16 April – the day the initial lockdown was extended – we carried an electric piano, plugs and wires, a mighty extension lead and an amplifier (disinfected and provided by our musician neighbour) and set up on the corner where three roads meet. It's quiet anyway, an avenue of glorious high trees and leylandii: horse chestnuts with their white candles proud, the first stirrings of colour on the copper beech, sycamore with its nesting rooks, mischievous lime and acacia. At dusk, it's a shimmering green and silver world and now the hum of traffic has gone there is only birdsong.

At seven fifty, I push Granny Rosie out from our garden and along the pavement. My mother-in-law, rising ninety, is dressed up and pitch perfect and ready to play when this evening's #ClapforOurCarers begins. My son helps position the wheelchair on the grass verge at the keyboard; my daughter sorts out her trailing wires and takes a picture; my husband checks the power. Rosie's fingers hover, she turns down her hearing aid, puts her right foot on the loud pedal and she's off. It's a wartime playlist – staples of the entertainment troupe, The Old Timers, that Rosie and my beloved Ma belonged to back in the day: 'Pack Up Your Troubles', 'Wish Me Luck', 'Bless 'Em All', 'Run Rabbit Run', 'Somewhere Over the Rainbow'. Many of the people who live in this patch of north Chichester remember these songs from their childhood. Younger families and those with grown-up children come 'home' like mine for the lockdown recognise them from the theatre, the big screen or the small.

The sun begins to set, a wave of clapping begins. Everyone has a reason to be doing this. It's not abstract, an obligation, but a profound desire to acknowledge a personal debt of care. For me too. For the past twelve years, I've been a carer in this house on the corner where three generations lived together – my husband, daughter and son; my mother and father; my mother-in-law. The NHS is part of our daily lives – the waiting room of our local doctors' surgery, the treatment and rehabilitation centres, A&E, the outpatients departments and the wards of St Richard's Hospital. My gentle and gentlemanly father, kind and principled, who lived with Parkinson's for many years until his death in May 2011. My vivacious and elegant mother, who was performing on stage with 'The Old Timers' in fishnet stockings and red mini skirt two days before she died suddenly in 2014 on the shortest and darkest day of the year. My magnificent mother-in-law, frustrated by her declining mobility and increasing dependence – she was a keen cyclist and horsewoman in her day – but still full of enthusiasm and curiosity. Known to everyone as Granny Rosie, she's something of a local legend . . .

There's a certain quality of silence in the outpatients corner of a hospital on a Saturday afternoon. No parking wars, the bustle of the week gone – just those with treatments booked and the incredible and patient staff. There's a certain quality of kindness in the cancer clinic, the COPD clinic, the day surgery, the reassuring competence of the reception staff. There's a certain quality of patience as the healthcare professionals step out into the waiting area for their next appointment, all eyes rising as people listen for their names to be called. The reassuring smile when nurse and patient connect before vanishing into an endless corridor behind the swing doors. The calm that then settles again, the air that stills again. Waiting, again.

Thanks to the NHS, my father could die in his own bed, as he wanted, surrounded by the people he loved and who loved him. He left his life as himself and with dignity. Thanks to the NHS, my mother left her life as she lived it, beautiful and witty and as herself, on a hospital ward where she felt cared for and loved. Thanks to the NHS, Granny Rosie continues to live her life as the woman she's always been. In a wheelchair now – and she feels the confinement – but her days are filled with jigsaws and knitting, G&Ts and whiskey macs, crosswords and Scrabble and books and good health for her years. Thanks to the NHS, Rosie was able to be outside in the road on that Thursday evening in April to say thank you in her own way.

When the clapping was over, Rosie began her last song – 'We'll Meet Again'. Of course. When I looked around, I realised there were now maybe eighty neighbours lining the avenues like a posed photograph – all observing social distancing, so sticking to their family groups or standing alone. Separate but, for the length of the song, together. People started to sing, hesitantly at first, then louder as the chorus came round again. When I turned around, I saw a mother and daughter cycling past had stopped to listen. A car pulled over and a nurse in her blue uniform got out. A final encore for her and an old couple danced, in an old-fashioned step, beneath the lime tree. Not a dry eye in the house.

Thank you, NHS.

Richard Mosse (30 May 1920–18 May 2011)
Barbara Mosse (15 September 1931–21 December 2014)
Granny Rosie (2 November 1930– going strong . . .)

FEARNE COTTON

Those Dressed in Blue

Phil and Sylvia Savage, the most wonderfully spirited pair,
Sylvia with her raspy laugh and Phil with his lack of hair.
He was a real practical joker, always talking to shop
mannequins,
Or picking me up unexpectedly to plop me in a shopping
cart bin.
We'd roll our eyes and laugh till we ached, yet he always
laughed the most,
His florist was full of rainbow and scent, his blooms his
biggest boast.
She was comfort and the biggest hug, with a catalogue
always in hand,
They'd theatrically bicker for our pleasure, yet side by side
they'd stand.
Phil adored being her husband and Sylvia thrived being
his wife,
My maternal grandparents, the Savages, such an important
part of my life.

*

Sylvia was first to be held by those loyally dressed in blue,
They cared and helped and bathed and spoke and did all they could do.
Her lungs drew shallow, clattering breaths and her hair came loose in hands,
Refusing a wig, a turban preferred to cover the last few strands.
Her laugh remained as comforting, and her nails still lacquered red,
Even during those last precious days, lying in her hospital bed.
Stories came thick and fast from the past, was she here or was she there?
Conflated narratives crossing paths, and then an empty wide-eyed stare.
Then once again recalling detail in a letter she would never send,
Yet those in blue came and went, right to the bitter end.

Phil's heart broke along with ours, but his was split in two,
To watch dear Sylv fade by his side, stood along with the rows of blue.
We tried to comfort, we bought him cats, but nothing could take her place,
His beautiful, cheeky, mischievous smile would rarely appear on his face.
His lungs started to mimic his dear lost wife's, which made him cough and wheeze,

He tried fishing for tiddlers and painting again but nothing would quite appease.
It was Christmas Day and those in blue had left us to give Phil a kiss,
They dealt well with his jokes and unusual ways and the odd acerbic hiss.
Much like Sylv his mind wandered and waved, stories rearing up from the past,
Glossed-over eyes and a fading smile yet his mischief till the end would last.
He left stories for us to laugh about and pretty paintings of oceans deep,
His last days peacefully lived out and then a surrender, he slipped away in his sleep.
Laughs never forgotten and shiny blue eyes, that are held in the heart so tight,
Thankful memories of those who cared and helped lead them both to the light.

Let the clapping never end for them, it is the least we can do,
To show our gratitude and love to all those loyally dressed in blue.

SUE PERKINS

It was the tail end of March 2018 and I was at home doing a spot of spring cleaning. It was the turn of the upstairs windows, which had acquired a notable layer of grime over the winter period. There was nothing out of the ordinary in my preparations; I placed the ladder against the wall at the top of the stairs, made it safe and then took off my clothes. I, like most people, prefer to clean naked – I find it simpler. I don't like to smear a newly scrubbed pane of glass with a cuff or sleeve and mess up my endeavours.

I always like to start at the top and work down. Makes sense. I stretched up as high as could, reaching for the top of the frame, and it was then I lost my balance. The ladder started to sway, gently at first, then violently from side to side – and, before I knew it, I was catapulted down the stairs.

As for what happened next, I can only submit to the laws of physics. Newton's First Law, to be precise. This states that an object at rest stays at rest unless acted upon by an unbalanced force. The unbalanced force was, of course, me – hurtling down the stairs. The object at rest was a hoover.

I'm sure you can imagine the discomfort involved in hitting a cleaning implement at speed, and at such an unusually acute angle too. Added to which I was naked, making my body even more vulnerable to assault.

I'm not sure if it was during the initial impact that the soft-brush attachment ended up where it did. Sadly, my memory of the event is a little hazy, due to the trauma. What I feel, is that if classical mechanics can't explain the whole story, then the answer must be found in the world of quantum, where spontaneous movement is entirely possible. Under these conditions, once I'd hit the hoover and the flex flew into the air, the attachment would be able to find its own way into that most dark and secretive of human corners.

Anyway, I told all of this to the nice man at The Royal Free Hospital and he appeared quite satisfied with my explanation. So much so, he brought in all his colleagues to hear it, over and over again. The ward was positively heaving by the end of the evening!

I cannot thank the NHS enough. Thanks to all its dedicated physicians and carers I am now attachment-free and can get back to cleaning those windows.

LEE CHILD

My only significant contact with the NHS started with a heavy leather football. My grandma bought it for me, brand-new, in the spring of 1962, when I was seven years old. It was an old-fashioned two-panel design, with a pink rubber inflation nipple nestled in an opening that did up with a square-sectioned leather lace, like you see today in a Sperry boat shoe. The first day I had the ball, I took it down to a patch of grass behind the Beauchamp Avenue shops, where I found a bunch of friends for a game. But that location was unsatisfactory, because the grass sloped at an angle of about thirty degrees. So we walked on down the hill, across the railway at Perry Barr, to a better spot. There we played all day, against all comers, shirts and skins, with the traditional piles of jumpers as goalposts. We quit when the light went and I walked all the way home. After supper, I started to feel lousy. I felt hot and achy, and I had a headache. Which was unusual for me. I was normally a healthy kid, solid and robust, who healed fast and was never laid up.

At this point I should introduce my mother. She was thirty-five at the time, the mother of three sons so far, and a hypochondriac, partly – I now realise – to get attention and to indulge a persistent martyrdom fantasy, but mostly because she loved doctors and surgeons and consultants in a giddy, girlish way. They were her movie stars. Her ambition was for us all to join the medical profession. (None of us did.) In the meantime, she took every opportunity to associate with her heroes in any way she could, either on her own behalf – or on ours. I hesitate to say Munchausen's, and certainly not Munchausen's-by-proxy, but any chance at all to be involved in the medical world was to be immediately seized upon, bright-eyed and eager and enthusiastic. Like phoning Clark Gable or Omar Sharif and having them show up at your house forty minutes later.

I remember the doctor well. He had sallow skin and a pattern of darker freckles on his face laid out as exactly as tribal markings. Initially he thought my aches and soreness were caused by a long walk to and from an eight-hour football game. He thought my headache was caused by repeatedly heading the heavy leather ball. Privately I agreed. I had scored some fine goals that day. But my mother, who had consulted her *Reader's Digest Family Health* book, wanted to be sure. She wanted to explore alternative diagnoses. Could it be rheumatic fever? Should I be placed on bed rest? Should I be in the hospital for observation?

The doctor responded scientifically, by saying – part reluctantly, part conscientiously – that he supposed it couldn't be ruled out. The result was I ended up in a bed on a ward in the Birmingham Children's Hospital. The next morning I felt absolutely fine. But I was kept there four whole weeks. I loved every second of every one of them.

It was a long ward, with beds on either side, possibly twenty-four of them, but only twelve or so occupied, all at the end away from the corridor. The patients were half and half boys and girls, all about my age. Nothing much seemed to be wrong with them. We could listen to the BBC Light Programme from a device built into the bedhead. I remember hearing Chubby Checker's 'Let's Twist Again' over and over. It was just out in the UK. Thus began a vital month of education – or re-education. Immediately I realised that my family – any kid's default definition of normal – was in fact not very. I suddenly understood that there were all kinds of different people in the world. I claim no deprivations or cruelties in my childhood, but it became instantly clear that I was trapped in a narrow, gray, repressed environment, desperately striving to join the respectable middle class. Now I was dumped into an entirely new family, made up of a dozen random Brummie kids, each with their own inherited beliefs and habits and norms. My eyes were opened, in all kinds of ways.

For instance, having only brothers, I knew little of girls. Certainly I had never seen one naked. That lacuna was quickly closed by the girl in the bed next to me. Her name was Jill. We cooperated. Show me yours and I'll show you mine. After our mutual curiosity was satisfied, we talked about all kinds of stuff, back and forth, widening into a yelled conversation involving the whole ward. Every day was full of chatter and stories and pop music on the wireless. I hated the visiting hour. We all had to quiet down and look suitably grave, me especially, to act my part. Then the parents would leave, and we would start up again.

Most of all I learned about women. My mother and grandmothers and all our approved neighbours cleaved to be respectable Edwardian-era middle-class stay-at-home housewives. The nursing staff proved there was another possibility. This was still the old starched-uniform era, with a rigid and fearsome hierarchy stretching downward from the always-capitalised Matron, through the always-capitalised Sisters, to the nurses themselves, who I thought impossibly glamorous and competent and bustling. We gave them a hard time. (Although now I guess they were happy we were happy.) Our beds were old iron hospital cots, with feet on one end and wheels on the other. We learned that if we grabbed the frame and braced ourselves in a certain way, and jumped and jerked, we could scoot the beds around the floor. We had night-time races up the centre aisle, sometimes out the door. It's where I learned to drive, really, at the age of seven.

It's where I learned a lot of things, at the age of seven. Four weeks in an NHS hospital, with nothing wrong with me, receiving no treatment, but for the first time in my life, I was left alone outside of my own suffocating, cult-like bubble. The right place at the right time. I still remember my growing up as before and after. I went home a different person.

MATT HAIG

The Inside of Hospitals

I can remember the second time I cut my head open as a kid. It was bad enough for me to have to go to hospital in the back of an ambulance. I had always been fascinated by hospitals – by what happened inside them – as if it was something as mysterious as the inside of a Tardis. I had a big gash on my forehead and blood leaked over my face as if someone had switched on a tap. But I can't remember the pain. The only thing I remember was the nurse holding my hand and asking in soothing tones about what I wanted for Christmas as the stitches went into my skin.

'Table football,' I winced. 'And a Liverpool kit. That's what I want.'

Within a year, I would be nine years old and would arrive at the realisation that I didn't really like football as much as I thought I did, but right then I had successfully convinced myself that everything was better if it had 'football' in front of it. The nurse smiled and asked my mum how likely that was, and by the time my mum had answered the stitches were done.

I can remember the anaesthetist, three years later, when I was eleven, who chatted to me before an operation. I told her I hadn't dared tell any of my friends I had to have it done as it was quite embarrassing to have, you know, your foreskin chopped off. And she smiled with comforting irreverence and said she didn't know as she had never had her foreskin chopped off.

I can remember visiting my eighty-four-year-old nan, night after night, following her last surgery as she lay slowly dying of breast cancer in a ward on the third floor of Newark Hospital's austerely Victorian building. I can remember my nan's bittersweet chuckle with her kind and jovial doctor as she asked if he was flirting with her.

I can remember my other grandmother, my dad's mum, in a hospital bed in Sussex, dying of so many numerous ailments it was hard to keep count. I remember her being more worried than my other nan, more aware of her own fate, her eyes saucer-wide. The nurses sensed this too and chatted with her tenderly, as if they knew words themselves were a vital ointment.

I can remember being an idiot at a New Year's Eve party when I was sixteen years old and drinking far too much cheap cider and standing too close to an outside fire and the wind changing. I can remember explaining to the burns consultant – as he gazed down at the large, fresh, oozing, multi-shaded purple wound on my left thigh – that I had been wearing two pairs of trousers as it was a cold night and so I hadn't noticed the outer layer was on fire until too late. 'You must have drunk a lot of cider,' he told me, without too much judgement.

I can remember seeing that same burns consultant – Richard, he was called, like my dad – only a few months later, for an even more stupid injury. This time my mum had asked me to see if one of the hobs on the stove was hot and, instead of acting like a sentient Homo sapiens and checking the temperature on the dial, or at least hovering my hand above the heat, I actually placed said hand directly on the hob. And such was the heat, my brain didn't translate it as heat for a little while so my hand stayed there for a few seconds until it was too late and my entire right palm was in a state of hot, pulsating agony, blistering in real time as I ran it under the cold tap for an eternity and then walked around to the hospital with my mum, in my pyjamas. Richard knew what to do. He lathered my hand with some magic cream and placed it in a transparent glove which I had to wear for a week, before returning to see him so he could transfer my hand into a bandage. 'Now,' said Richard, poor bloke, smiling a tired smile, 'promise me you'll stay in on Bonfire Night, yeah?'

I can remember wandering around a maze of hospital corridors after the birth of my first child, dazed and euphoric and bewildered, eventually helped to the car park by an off-duty doctor.

I can remember my wife, Andrea, bleeding and nearly going into premature labour at twenty-three weeks with our second child, and the consultant obstetrician – after effectively saving my unborn daughter's life – gave me a stern look: 'Your wife must stay in bed for seventeen weeks and you will have to do everything, do you understand? You are her personal butler!'

'Yes. I understand.'

I can remember the kindness of the oncologist at the Queen Elizabeth Hospital in Gateshead who was treating my mother-in-law for ovarian cancer, and chatted to her about garden centres. The pleasant triviality of the conversation as welcome as rain on the savannah.

I can remember my dad, coming around after twenty-four hours of unconsciousness after nearly drowning in a lake, on so much morphine he sincerely thought he was in the afterlife.

I can remember my mum, two years ago, in the intensive care ward after having open heart surgery to replace her aortic valve. She'd also had morphine, but instead of it causing her to imagine she was in heaven, she had felt she was about to be experimented on in a sinister laboratory and was now feeling a bit embarrassed about how she may have acted in front of the ICU staff. Not that they seemed fazed in any way. In fact, they were lovely. One of the porters made sure the window opposite my mum had the blind opened fully, so that she could see the sunshine and sycamore trees beyond, which cheered Mum up immensely.

I can also remember last week, when Andrea chopped

the tip of her finger off with a kitchen knife and the blood wouldn't stop. We tried everything. We wrapped it up with kitchen roll. We grabbed a bag of frozen sweetcorn. We placed it over her head (the finger, not the sweetcorn). But nothing. An hour later and it was still bleeding, and we called 111 and – despite the fact we were one week into lockdown and hadn't left the cocoon of our house for anything more than to take the bins out or the terrier for a walk – we now had to go to A&E. With a bandage around her leaking crimson finger and a scarf around her face, Andrea went in to see the triage nurse. As the nurse cauterised Andrea's finger, Andrea asked if he was worried. 'About what?' the nurse asked.

'You know. *The virus.*'

'Of course. But you just get on with it, don't you?'

And I remember this A&E department was part of the same Sussex hospital my worried grandmother died in years ago. I remember how she had been comforted by the nurses who had sat beside her and held her hand and spoken and listened, alongside their other duties.

'They are ever so good,' she told me. 'They really are.'

And the nurse in the room had just waved it away as if it was nothing at all, before leaving to gently attend to another duty.

JOHNNY VEGAS

So there you were, a nurse, in a club. You were patient, funny, with the same kind of acceptable gift for the darker sides of humour. An equal ability to laugh at adversity without sacrificing empathy. I fancied you, I won't lie, but I also thought we had so much more in common.

We started off discussing awkward moments – me as a patient, or visitor, and you the administrator of care. I drank some more and wanted to prove to you that I wasn't just about joking around these moments, that I felt we had a kinship. I went on to explain how I deal with the prospect of death all the time. 'Oh, I had this shocking gig in Wakefield, died on my arse . . .' I then proceeded to tell you all about me, my routine, my dread of going to work at times.

You asked me politely, because I was already testing your social graces, 'But you must enjoy it, otherwise why would you bother?'

'Well . . . it's a living I suppose.' I thought by saying that I could retract all of my past fifteen minutes of egocentric nonsense.

You said your mates were waiting and wished me good luck with it all – the comedy and that.

I'm committed to comedy, much more than folk might realise. I've suffered for my art but been paid handsomely at times for trading in my passion for a stronger pension. I think of that night more than I should. Not because I'm weird and obsess over the misconception that a random chat meant she might have been 'the one'. I think of her before gigs, before telly spots, before having to be funny when I can't be arsed. I think of how she stormed every shift and put me to shame, made my parents smile through excruciating pain. How she had her own grievances but left them at the hospital doorstep and brought something more than I was capable of to every day she gigged.

I'm a people watcher. I'm fascinated by how and what motivates folks' functionings. Yet, sat in many wards, over many years of late, wishing for the best on behalf of those I hold dearest, I honest-to-God was able to switch off, stop trying to focus on the motivations or backstories of those who worked the wards and simply surrender to an unusual peace made up of gratitude and admiration.

I wish I could bump into that nurse again and just say 'thank you' in passing, praying that she might believe the sincerity experience has woven into it.

PAUL SINHA

On 26 July 1990, on what was a small, barely adequate television screen, I watched the first day of the first Test between England and India with my dad by my side. It had been an uneventful day's action thus far. Predictably enough, England's openers faced little to concern them from the tepid Indian bowling attack. Then Graham Gooch edged one. The collective hearts of the Tebbit-ignoring father and son leapt with joy.

The joy was momentary.

The wicket keeper, Kiran More, dropped a chance that was easier than the questions on *Tipping Point*. 'Out!' exclaimed my dad from his hospital bed.

'I'm afraid not, Dad,' I replied, 'More's actually dropped it.'

'What an idiot,' my dad continued, with the characteristic frustration and sadness that India's touring team regularly seemed to evoke. 'Just mark my words, Gooch will get a double-century – or even a triple.'

'Well, since you're coming home today, at least we'll watch the rest of the match on a proper TV.'

'Good point. What a month.'

'Yes, Dad. What a month.'

And as we pondered India's failure to capitalise on the easiest opportunity imaginable, a cardiac nurse at King's College Hospital, Camberwell, came in to take my dad's blood pressure and give him clear discharge information. And I, soon to be a second-year medical student, nodded sagely and pretended to understand. It really had been quite a month.

A few weeks before, I'd been on the cusp of the biggest adventure of my life. I'd just passed my first-year medical exams at the third time of asking. My dad, fuelled by massive generosity and overwhelming relief, had decided that I needed a holiday. That holiday was to utilise a Delta Airlines unlimited-flights, one-month pass and Have Fun in the USA.

I had spent sleepless nights trying to formulate an itinerary and at last I'd solved the conundrum. New York to Miami to Key West to New Orleans to Los Angeles to San Francisco to Chicago to New York. (Incidentally, it had not occurred to me at the time that the fact that I was not yet twenty-one would be problematic. This trip was never going to realise *all* the ambitions of a travelling twenty-year-old.)

I'd only been in New York four days when the family friends with whom I'd been staying had news for me.

Tomorrow's flight was not going to Miami; I'd already been booked on a flight to London. My dad had had what was described as 'a major cardiac event' and was currently in Intensive Care at King's College Hospital. I had a mum and a sister – and, of course, a dad – who really needed me back home.

Most of the next four weeks were something of a blur. It had been quite clear that my dad had been boasting about 'my son, the medical student', given the degree to which the calm and courteously professional nursing and medical staff insisted on talking to me on the misjudged understanding that I was something of an expert on cardiology.

As a matter of urgency, my dad needed a bypass operation. And my hazy memory places these rather dramatic six hours as starting on a Monday or Tuesday afternoon. Of course, we should've stayed at home and waited for news. But it was the summer holidays and, quite frankly, my mum, my sister and I had very little to do. After all, what could be more fun than sitting in a relatives' room, waiting to find out if your dad was going to live or die?

What I do remember was the patient kindness of auxiliary staff, who may have lacked the glittering academic achievements of their medical colleagues but certainly lacked for nothing in intuitive empathy. And that day, they needed to be at the top of their game. While emerging from general anaesthesia at the end of a lengthy cardiac bypass operation, my dad had a cardiac arrest.

Looking back, the next forty-five minutes were the most significant forty-five minutes of my life. Years later, as a junior hospital doctor, I would take part in many such attempts to resuscitate patients. The results were not always spectacular.

I never met any members of the arrest team that day and I've never met any of them since. What I do know is, at the age of fifty, my dad was on the brink of death. And thanks to their skills, hard work and determination, that death didn't happen on their watch. My dad, thirty years on, is still loving life.

I have no idea where I'd be without that cardiac arrest team. Everything that I've achieved has been fuelled by the emotional and financial generosity of my parents. Medical education aside, this was my first major NHS experience and it defined who I am today.

It's something of a miracle to still be alive thirty years after a cardiac arrest. The problems have never gone away and, in 2020, my dad had what felt like his one billionth cardiac operation. Every new presentation has been greeted with a benign smile and a 'Come on then, Dr Sinha, let's sort you out' attitude from a rotating roster of skilled professionals.

On 14 December 2019, I got married. I got married to a man. In 1990, if someone had told me that, in thirty years' time, this would happen, and – rather than be upset or irate, or dead – my dad would be drinking whisky all day with a massive smile on his face and a life-affirming determination to meet as many of my friends as humanly possible, it's fair to say that, at the very least, I'd have been mildly sceptical.

As I write these words, every day I've been video-calling my mum and dad, making sure that they're currently emotionally strong enough for the coronavirus lockdown. Strong enough? They are, at this stage, used to peril. And I don't think it has ever slipped my dad's mind. He's ready for whatever the global pandemic throws at him. Because he knows that, in July 1990, the NHS played an absolute blinder – enabling him to become a grandad, a father-in-law to two men, and husband for what is now fifty-two years.

I now have my own NHS issues, thanks to a recent diagnosis of Parkinson's disease. When I go to my consultations, I take the time to breathe in and look around. This may not be my line of work any more, but it is the NHS that has made sure that my Parkinson's is a fight that I embark on with the support of both parents. And for that, many thanks.

BARONESS TANNI GREY-THOMPSON

Dear NHS,

Thank you! Without you I would be dead and I wouldn't have had the chance to have the life that I now have. I wouldn't have had an education, become an athlete or be a mother. If I had survived, my life wouldn't be the one that I have now. Many of my happy memories of you are wonderfully kind and funny comments that have been made in some quite serious circumstances.

I was born with spina bifida and didn't need a lot of care at first, but then I had operations at seven to have a good look at what a mess my spine was in; at thirteen to attach metals rods to my spine to stop it collapsing further and then at nineteen when they snapped when I fell off a rope in a training session. Sorry, that last one was my fault, but you didn't judge me and one medic jokingly said (when I was in a reasonable amount of pain), 'Well, you probably shouldn't do that again!' A sensible piece of advice and matched the thoughts, if not the tone, of my mum's response to finding out what I had done.

There are so many people in the NHS I should thank. To those who, when I was a child, spoke to me and not my parents (they strongly supported this), and to my surgeon who, when I was twelve, told me: 'You need to understand this because it is you who will be going through the operation.' That set an important tone for the rest of my life.

To the lovely team at the University Hospital of Wales, who delivered my beautiful (she will hate me for this) now eighteen-year-old daughter by C-section and didn't think I was crazy when I said I had to be back in training really quickly to represent Wales at the Commonwealth Games. Thank you to the team who didn't laugh at my Hull-born husband who wanted me to go into theatre with a small sticking plaster on my shoulder with some earth from Yorkshire on it (yes he did bring it down, so she was born on 'Yorkshire soil'); it has given me a fine story to tell.

To the doctor who, when I burnt my knee on a radiator and treated it myself for a couple of days before finally going to hospital (I had done it before, and I wasn't near home), looked at my felt-tipped knee with its various colours and lines to show different points of the burn, and just said, 'You've done this before.'

I have been privileged to travel the world and I have seen that lots of other people don't have the medical support we do. I am lucky that I don't have to worry about not being treated because I don't have the right insurance, or based on the ability of my family to pay.

To all those who, day in and day out, are on the front line, dealing with things that most people can't imagine: thank you for being there. We have always valued what you do – the weekly clapping just shows it – and 'when we are on the other side of this' (if I had a pound every time I heard that I would be rich) I hope we can show you in other ways how much you mean to us.

RENI EDDO-LODGE

Haemoglobin is a protein in your red blood cells that transports oxygen around your body. Understanding that we need oxygen to survive is primary school-level stuff, but until I was admitted to hospital, I didn't fully grasp that it was that little-known protein doing the heavy lifting in keeping me alive.

There's a morning in my life I'll never forget. The events are burned into my retinas. I'd taken a blood test a week earlier and my GP, on receiving the results, phoned. There was urgency in her voice. 'You need to go to A&E. Now.'

I was twenty-five. I'd just moved house and had had to push myself physically to get boxes up the stairs. I was tired. I struggled to shake off the morning grog, no matter how much coffee I drank. If I stood up too quickly, my vision would be blighted by black splotches. Sometimes I couldn't properly form my thoughts into sentences. My usual twenty-minute walk to Sainsburys had been replaced by a bus journey that would get me there in half the time, because I was too tired to walk. I'd sit on the weaving, round-the-houses, single-decker bus alongside pensioners, toddlers and parents.

'When you get there,' my GP said, 'tell them about your haemoglobin levels.' I can't remember the exact number she told me, but it was dangerously low. Six-point-something. Far lower than the amount a healthy human needs to run on.

I was in disbelief. 'Can't I just take some iron tablets?' I asked. 'No,' she said. 'You need to go now.'

I can't remember how I got to the hospital. Maybe I splashed out on an Uber. More likely, because money was tight, I got on the same little bus that took me to and from the supermarket. If I stayed on it long enough, it would deliver me to the hospital entrance. In A&E, I sat in a plastic chair opposite a staff member behind a window and explained my situation. They gave me what looked like a raffle ticket and I sat in the waiting room, my tiredness loosening my limbs.

Someone came to get me with a wheelchair in tow. It was a long walk to the women's ward, they said. I sat down in the chair and was zoomed along corridors. My shoulders sagged and my body folded in half. I began to sob. The staff member wheeling me tried to offer some comfort. What I would have said at the time, if I'd had the energy, was, 'What the hell is happening to me?'

It has taken me a long time to accept that I am chronically ill. I've long fostered a (fairly damaging) core belief that the only person I can rely on is myself. I've dragged myself to the edge too many times and pushed myself too hard in the belief that there would be no one else to help me. A sudden admittance to hospital disrupted that individualist notion.

I later learned that the effects of low haemoglobin – when you don't have enough iron in your body to effectively transport oxygen on your red blood cells – are not unlike carbon monoxide poisoning. It's a silent killer. There's no extreme pain, instead you just – slow down. Carbon monoxide latches onto the haemoglobin on your red blood cells so effectively that the oxygen can't get a look-in. If it's not caught quickly, you'll drift off and you won't wake up.

My low haemoglobin levels were due to excessive blood loss – heavy periods that have wrecked five days of my life a month since puberty. I remember falling asleep in the middle of the day at university. I once tried to put a metal plate in the microwave; my boyfriend stopped me. The blood loss is so severe that during the rest of the month, my body can't keep up with last month's loss before the new month's loss begins.

In the women's ward, everyone in the other beds was at least double my age. Wires were hooked up into the vein of my left arm and I was told I was having a blood transfusion. I thanked every god in the universe that I was born in a country in which this health catastrophe wouldn't plunge me into bankruptcy.

The sounds of the women's ward were disorientating. In the bed opposite me, a very old lady muttered 'Please help me' at three-minute intervals and fought with any staff member who heeded her call. Next to her was a middle-aged woman with a thick plastic bag of dark urine hanging off the side of her bed. She wailed through the night. In the small hours, I asked her to keep it down so I could get some sleep. 'I'm sorry, my darling,' she said. 'I'm in a lot of pain.' I felt terrible for chastising her.

Overnight, three litres of fresh blood were drip-fed into my veins. At one point, I had to go to the toilet, hooked up to my drip. A nurse helped me take the blood with me.

The next day in the hospital pharmacy, a friendly pharmacist let me take home a bulky paper bag full of medicines for free. There were dozens of pills. Two types of pills to stop the bleeding when it arrived. Pills to top up the iron levels in my newly acquired blood.

Days later, lying on the sofa in recovery, I was awash with a sense of gratitude. I marvelled at the fact the bagged blood was there when I needed it. I wondered who my blood donors might be. Did we live similar lives? Did we have different political views? I was in awe of the altruism. I urged everyone around me to give blood if they could and, since I can't give blood, I signed up to be a posthumous organ donor. The card is still in my purse.

While it's not entirely dissipated, the core belief that I could only ever rely on myself began to erode after my hospital trip. I needed an NHS to lean on and an anonymous donor who shared my blood type to donate just in time when I needed it. I needed a smiling in-house pharmacist to sort me out with the medicines I couldn't afford at the time. I needed a nurse to help me get to the toilet. I needed a community of people who had dedicated their lives to care.

I still get bouts of crashing tiredness after physical activity. A long lockdown walk to the pet shop had me wiped out for the rest of the day. Iron supplements are part of my daily routine. Coronavirus has severely delayed the scheduled surgery I was due to have that would have gone some way in resolving my underlying heavy bleeding problem.

I'm still managing my iron-deficiency anaemia. But thanks to that blood transfusion, the NHS and the kindness of many strangers, I'm still alive.

PROFESSOR GREEN

Dear NHS,

Shortly after you delivered me into this world (the nurse handed me to my grandmother who would later become my legal guardian – not sure if it was intuition or just because my mum was sixteen and, despite having just given birth, looked too young to be a parent herself) I ended up back in your care and in need of surgery.

At six weeks old, one of your surgeons performed a pyloromyotomy and again gave me the gift of life.

But as grateful as I always have been for your care, I don't like hospitals.

When I was thirteen, you cared for my great-grandmother as she slipped away, a huge life event for me. I don't know how you find the strength to get up and go to work every day not knowing what you're going to encounter – not just the loss of life and suffering, but the anger, the abuse (I've ended up in Homerton A&E on a weekend and witnessed it first-hand!), the people who don't truly appreciate just how much you give in what you do for us mere mortals.

Over the years I've spent some solid time with you guys:

Two weeks for glandular fever when I was sixteen (one of your nurses asked my then-girlfriend not to sit on the bed and later accused me of cheating on her as she didn't have glandular fever – I don't hold it against him, I'm sure he was just having a bad day).

Two weeks for what looked like colitis but thankfully turned out to be food poisoning (campylobacter – I was happy to leave with the entire length of my large intestine and without a poo-bag).

And then again at twenty-four, when I was stabbed. After five hours in surgery, thirty-eight internal stitches (no idea how many external) and with the possibility of me not waking up or being disabled due to nerve damage, you brought me round and asked me to shrug my shoulders to make sure I still had the use of my arms. I did. Phew. When I glanced at my reflection for the first time and saw the wound, I thanked the surgeon for putting my two-week-old 'Lucky' tattoo back together; he thanked me – apparently it made his job easier as it helped him work out what went where!

Some ten years later I did some work to raise money for the Trauma Research Unit at the Royal London Hospital and visited other youths you'd saved from the fate I nearly met on the very same ward I woke up in. A small token of appreciation in the scheme of things.

I've seen you a couple of times since then as well. I was squashed between two cars on 23 May 2013 – four years to the day I was stabbed, actually! Thanks for the gas and air, and the picture – I always laugh at how my hair somehow remained perfect throughout the whole ordeal.

More recently, I saw you after a seizure during which I fell and smashed my head – I'm haemophilia B / factor VII deficient so you rushed me through for a scan to make sure there wasn't a bleed, which there wasn't – news you delivered much to my delight, though in your next breath you broke it to me that, despite not having a bleed on my brain, I had in fact fractured my neck. I've since made a full recovery.

I'm not sure if I should end this with an apology for all the time you've had to put up with me or a thank you for always coming through, despite being overworked, underpaid and (by some) under-appreciated. I will always rally for you lot because, unlike the many false idols (myself included), you are heroes. Real-life ones.

I'm sorry it took a global pandemic to highlight just how incredible what you do on a daily basis is.

Thank you NHS,

Lots of love,

Stephen x

WILLIAM BOYD

Three Doctors, a Dentist and a Nurse

Back in 1971, during my first year at university (in Glasgow), aged nineteen, I began to suffer from debilitating headaches. They would come and go at random, while I was eating lunch, sitting in a lecture theatre, walking back to my hall of residence. They were mainly on the right-hand side of my head and they were so severe that I had to close my eyes, clench my fists, stop whatever I was doing and wait for them to pass.

Of course, I thought I had a brain tumour.

I went to the university's medical officer and he suspected I was suffering from chronic sinusitis. He arranged for me to have an appointment with an ear, nose and throat consultant at the nearby Royal Infirmary, a huge, sprawling Victorian hospital. It was my first experience of the NHS, having been born and raised in Africa and educated at a boarding school. I was a healthy young fellow but these headaches had rocked my sanguine insouciance about my well-being.

The ENT consultant was a bit baffled but he conceded that it might have something to do with an inflamed or congested sinus and he said it would be a good idea, anyway, to have an antral lavage – in other words, to have my sinuses hosed out. Anything to get rid of these headaches, I thought. And an appointment was swiftly booked.

The headaches continued and I counted down the days to my appointment. I learned that a cannula would be inserted into the maxillary sinus, attached to a hose and a saline solution would wash the sinus out. I would need a local anaesthetic, it transpired, and here's where things went wrong.

The anaesthetic that was being wiped around my nasal cavity with a swab had a powerful unforeseen effect on me, for some reason. The room went red, then yellow, then blue. It was as if I was tripping on LSD. I went deathly pale and the room began to sway. The nurse who was applying the anaesthetic stopped at once and she went to fetch me a cup of tea. I came round very quickly and apologised. I said I was ready to give it another go. But she said: you know what? I think we should postpone this procedure. I agreed with some reluctance – I desperately wanted rid of these headaches – but she was insistent.

When I returned to my hall of residence there was, coincidentally, a letter from my father. My father was a doctor, also. He was a specialist in tropical medicine and spent his working life in West Africa. I had written to him listing my headache symptoms and now in his letter he – strangely, I thought – suggested I go and see a dentist and have my teeth x-rayed.

I went to a nearby NHS dentist who duly x-rayed me. It turned out I had a huge abscess under my rear right molar. It wasn't giving me toothache, yet – that would have arrived in due course – but the purulence was infecting the nerves that ran up the side of my neck, causing these intense headaches. The dentist removed my tooth and drained the abscess and the headaches went away forever.

I was, of course, beyond grateful to the NHS doctors who had seen me, and to my father for his diagnosis at a distance and to the NHS dentist who had finally solved the problem. But, curiously, it's the nurse who *didn't* perform the antral lavage who earns my lasting gratitude. She saw the state I was in and unilaterally decided to postpone the procedure. It was an act of caring that involved withholding care. I didn't need an antral lavage as it turned out, not that she would have known, but that understanding of my distress – that empathy – contributed crucially to the curing of my headaches and allowed the dental intervention. It is a nice paradox, but that nurse's instinct not to do what she was there to do explains my undying devotion to the wonderful institution that is the National Health service.

VICTORIA HISLOP

It Means the World ...

For every moment of my life, the NHS has been there:

- It kept me in the world when I had meningitis.
- It brought into the world our two beautiful children.
- It gave my mother a long life in this world, curing her of cancers and heart problems and many other conditions. Last month, aged ninety-two, she passed away, in peace.

Thank you, NHS, you mean the world to me.

PHIL WANG

Have you ever stumped a doctor? It's rarely a good sign. The last thing you want to hear as a medical professional investigates your gaping mouth is, 'Well, that's new.' But that's just how special Olga was.

The main point of this story, as will be the case with many in this book, is that the NHS is great and kind; our nation's proudest achievement. The second point is, don't bother your ulcers. Really, don't. This isn't one of those pearl-clutching warnings of the overcautious, like 'don't eat too much cured meat' or 'don't get in with the mob'. This one is real. And I ignored it to my detriment at some point in 2009.

I grew a delectable little ulcer on the inside of my bottom lip. A cute little bump with a tiny white head poking out the top. Perfect for all-day sensual licking and the occasional cheeky nibble.

I nibbled on that thing enough to give it a life of its own and to develop feelings. Angry feelings, it turned out.

Because it soon swelled to six times its original size and transformed into something new entirely. It Hulked out, lost the adorable nibble tag and turned into a vengeful, painful, purple orb. An absolute shiner. The kind of marble children could fight over.

The super-ulcer became so large and so sinister that my sister gave it its own name: Olga. Sorry to any Olgas, but I'm afraid it was spot on. And to be frank, a quick glance at European history will confirm that Scandinavian names are quite appropriate for anything large and invasive.

Olga sat on my bottom lip, pushing against the top. Distorting my mouth so that I had a permanent grumpy smirk. Like Popeye had started taking it easy at the gym. Olga would fill with fluid until she could take no more, then burst her own banks, releasing a sticky midpoint between mucus and saliva that formed translucent vertical tendrils whenever I parted my lips, like I was being silenced by Agent Smith in *The Matrix*. She would then heal over and start slowly refilling again. And so the futile cycle went – an icky Sisyphus.

The GP googled my symptoms ('lip abscess sticky fluid bad odour potentially unrelated') and found Olga was most likely a mucocele – a mucous cyst that develops when salivary glands become plugged. They are normally painless and temporary, but only if you leave them alone. I, however, had been chewing on Olga like tobacco, and now she was going to make me pay, like tobacco. Olga wasn't going anywhere.

I was booked in to have her removed, and spent the intervening days saying my goodbyes – watching films; having dinner together. When the appointment came, I was met at the oral surgery by a nurse and surgeon who were calm, kind and smiling. It has always amazed me how NHS staff are able to put you at ease with the implied unremarkable routineness of your procedure without making you feel any less welcome or important. It's a difficult balance to strike, and yet it is achieved 99.9 per cent of the time (I am keeping in mind the time I coughed open-mouthed directly at my university GP, which made him instantly catch a very bad case of shouting at me).

The worst part of the experience was receiving the local anaesthetic in my lip. Compared to 'general', 'local' anaesthetic had always sounded relatively pleasant and benign to me. Like an artisan bakery or a parish meeting. So it's strange that it should be the scarier of the two options, and that it meant I had to watch and feel a needle stab the softest part of my face and then impregnate it with fluid.

After that, the rest was surprisingly tolerable. The surgeon sliced open my lip on the inside (to keep the scar invisible, the legend) and began sawing around Olga like a bone in a tough steak. This was one of the most peculiar sensations of my life – I was completely free of pain while something was being cut out of my head.

He eventually severed Olga's final grip and lifted her in front of me. She looked awful naked. Pink and glistening and bloated and alien. Which made what the surgeon said next all the more surprising.

'Can I keep this?'

'What?' I said.

'It's just . . . I've never seen anything like it before.'

'Um. Yeah sure.' His curiosity charmed me. 'Her name is Olga.'

'What?'

'Nothing.'

My lip was sewn up, my mouth filled with cotton balls and I was sent on my way. Easy as that. I thanked them both and walked out of the room, turning back to steal a final look at my old friend. The new man in her life was placing her delicately into a vial. To study her, understand her, learn from her and better serve his noble purpose. Or because he's just a fucking pervert or something.

MIRANDA HART

Camaraderie and Comedy

A hospital waiting room is not a place we choose to spend time in. It's not somewhere we roll up to for the 'good times' – unless you have a particular penchant for *The Reader's Digest* and dubious vending machines. They are places of boredom at best, but often a heady mix of palpable worry and ominous expectation. So, when we hear our names called for our turn into the unknown, how we are greeted and treated as we leave that menacing waiting room is crucial. We are in much need of a kind face and a reassuring word to counter the internal wobbliness.

One waiting room I remember in particular was in 2010 when two bizarre events collided in my life – an increased level of fame and the need to have an endoscopy and colonoscopy. If you want the layman's terms – a camera down my throat and a camera up my bum. You're welcome! Knowing that we were all there for one or both of these intrusions meant very little eye contact in the waiting room and a certain level of collective bottom wiggling on the plastic chairs. My name was called out. My turn for the unmentionable. Off I sloped, anxious, in dread. And I really was anxious, dreading both procedures. (Awful word – procedure. It's deeply suspicious; I don't like it!)

But within a few minutes of meeting the nurse, my unease was, well, easing and I found myself with an unexpected smile on my face. I was indeed greeted with an extraordinarily kind face that began to placate my distress of being in a situation I just simply did not want to be in. I don't know about you, but hospitals for me bring out a very raw, childlike fear, a feeling of wanting to escape, of being on the edge of not being able to cope at all and having to dig very deep into my resources to simply stay in the room without a massive tantrum. I am not at my best in hospitals. So, to find myself smiling was pretty extraordinary and unshakeable proof that it's the workers within these institutions that make the *entire* difference. Without their sacrifice and love, the necessary care would simply not be there for us in order that we don't escape, but get treated and stay well.

The reason for my smiling was perhaps a unique one – the nurse said that the ward had been excited to hear that I was coming in that day. She then proceeded to say that they all drew straws as to who was going to treat me. As I had just been informed that the first item on my treatment agenda was to have an enema (it's best you look that one up), I replied: 'Well don't tell me whether you got the short or the long straw, because both are *terrible* answers!' If she got the short straw I would remain constantly apologetic, and if she got the long straw then it indicated that she wanted to, umm, become familiar with my bottom. Oh god, are you still reading?! She laughed, I smiled and she continued to treat me with incredible gentleness, guiding me each step of the way with clarity of information and holding my hand when I requested it.

You might think I had preferential treatment because of the whole fame thing. And before I was put into a mild drug-induced state I had the same question. But I looked around and, not a bit of it, was the answer. All the nurses were showing the same level of attention to even the most cantankerous of patients. There was one who acted like a furious dowager duchess and I remember thinking that Maggie Smith could play her perfectly if they made a film of this time in my life – the drugs were now kicking in . . . (That's a film NO ONE WANTS TO SEE!)

This silly story is all to say thank you. Thank you, dear NHS. For each and every one of you within it that sacrifices so much to provide a unique skill essential to every one of us. We may not like hospitals, but we love you. It is your shared purpose, your togetherness in your uniquely pressured job, your support for each other, that gives you the strength to be the best you can be for us. That day in 2010 may have had a ward teeming with people in discomfort due to humbling and horrifying procedures (there's that word again), yet all I saw was a jolly camaraderie, which in my case produced some much-needed comedy. It's a smile between nurse and patient I will always be grateful for.

CATHERINE MAYER

Love can be as frictionless as silk.

I love, without restraint or qualification, Andy, my husband of three decades. I have nothing but love for my stepsister Sarah and my stepfather John; for Maurice, integral to my sister's life and our family for seventeen years; for close friends Sara and Barbara. I used to love the NHS unconditionally, as an ideal. Then, six times it failed me, six times over the past four years. In neglecting to save Andy and my stepsister and my stepfather and Maurice and Sara and Barbara, the NHS transfigured six vital, protean relationships into one-sided adoration, boundless but hopeless.

Alone in self-isolation, I mourn the exquisite messiness of living relationships. It is no comfort that my feelings for the NHS have come to incorporate some of that vanished complexity.

Sara erupts into the corridor, dragging her drip stand, a boat broken loose from its moorings. She pleads for morphine. The previous shift missed her prescribed dose and cancer grinds at her bones. The duty nurse shakes her head; she cannot override the automated system. Later the same day, a few floors above, my stepsister smiles in welcome, too weak to raise her head. She has asked me to visit, to receive her parting thoughts. I pull shut the curtains around her bay, lean close, but the bleep and clatter of the busy ward steal more than a few of her precious words.

My stepfather will survive several stints in the same London hospital before the last, when they hang a sign of a swan at his door to remind themselves to leave him, and us, undisturbed. That's the thing about the NHS: call bells routinely go unanswered, yet once the patient is beyond help, attention can seem overabundant. Six times I have observed this, the impulse to adjust the angle of the bed or essay a cheery conversation, as if such courtesies could make any difference to the dying. Barbara, at the hospice, yellowed and hollow, has not eaten for days. Even so, every mealtime, they trip into her room, singing out the menu.

All this I could forgive and do. I struggle with a sharper betrayal. Andy, admitted to intensive care with pneumonia, is expected to make a recovery. 'You're the healthiest person in ICU,' the doctors tell him. (The longer he's there, the better we comprehend how low that bar is.)

Five and a half thousand miles away, Wuhan is in lockdown. Reports seep through of European infections. Andy visited China recently, so the doctors decide to test him for the new virus. The test proves negative but

Andy's condition deteriorates. A tense narrative unfurls in coloured lines on the screen behind him. In those first days, I am a child, sounding out its first words phonetically. By day ten, I have learned to read the lines with ease and, anyway, I read faces. I understand that things are precarious.

So why do I allow the nurses to persuade me to go home? He is stable, they tell me. Get some rest. This will be a marathon, weeks if not months before he can be discharged. I withhold these prognoses from Andy. He still talks of touring with his band in the summer. From his hospital bed, he issues instructions for mixes of an album, like its creator, entering the final stages.

Sleep well, my sweetheart, I say. I'll be back early in the morning, and I am. As I text him from the hospital entrance – does he want a coffee? – a nurse calls. She says they have placed him in a medically induced coma, on a ventilator. All at once I am running, along hospital corridors, not waiting for that stupid lift, up the stairs, more stairs, using all the breath, all the breath Andy needs. No matter how fast I run, it will not be fast enough. He will never speak again, though he lives for another five days.

It is not their betrayal, those nurses. It is mine. I knew he was sinking but flinched from the knowledge. Now another piece of knowledge curls around the edges of every waking moment, making my eyes sting: that I squandered the opportunity to hold Andy's hand one last time and feel an answering squeeze, to talk with him rather than at him. (Believe me, I talk at him every day.)

No blame attaches to the NHS. Over four years and those strange and terrible fifteen days in ICU, my relationship with this extraordinary institution has been tested and sorely tried, emerging as a love based on intimacy. I have seen the flaws and failings but also the care. I have seen dedication, exhaustion, the toll that daily exposure to human tragedy exacts and the commitment to carry on regardless.

The critical care staff looking after Andy were exceptional, but they are not exceptions. They drew on specialists and their own deep expertise, deployed new treatments, explored every option to give all of their patients the best chance possible. I have seen similar excellence in other NHS hospitals, in specialist units and on general wards. I have seen bright successes against odds stacked by years of under-resourcing and bad policy. There are and will be many such successes.

I have seen that small courtesies might do nothing for the dying, but mean everything to those of us lucky enough to be with our loved ones when they die.

The end is near.

A consultant and I reach this decision on the final day of January; one more night and then a withdrawal of life support. Outside, protestors with candles keep vigil for the European Union. Inside, family and elective family take turns by Andy's bed. His bandmates curl on the window ledge. At dawn, someone brings in the first edition of a broadsheet; its Brexit headline is 'The Day We Said Goodbye'.

And so we will. First there are arrangements to make. Family and friends have dispensation to remain, despite their numbers. I choose a piece of music, not one of Andy's galvanising compositions, but a classical piece that he liked to hear as we drifted to sleep. It would be cruel to pull him back towards consciousness. Even so, I accept an offer from one of the nurses to shave him.

She does so tenderly, omitting only the section of his upper lip obscured by the ventilator's mouthpiece. He cannot die with a Hitler moustache, I protest. She agrees and prevails on a second nurse to hold the plastic out of the way so she can complete the job.

Small courtesies.

It is time now. Now there is no more time. I tell Andy, again and again, that I love him. I stroke his beautiful face.

Love can be as frictionless as a freshly shaved chin, as sharp as a razor.

ALEXANDER McCALL SMITH

Those Who Care

None of us remembers that first meeting –
Tumbling out into a very different world,
Into your receiving hands;
Blinking at the light, we breathed
The strangeness of oxygen
Saw the unfamiliar walls
Of the delivery room,
And the first ceiling we had ever
Looked at and wondered what it was.

That was the first thing
You did for us: welcomed us,
Ushered us into infancy
And childhood, and the years beyond.
We never thanked you
But do so now, rather late,
But with all the feeling

Of the long overdue
Expression of gratitude,
The tardy repayment
Of an ancient debt.

Since then, from time to time,
You have picked us up,
Dusted us down, bandaged
The occasional consequence
Of our failure to look
Where we're going
Or to behave quite as we might;
Tolerant, like all good
Members of your professions,
You said nothing, but did
What was necessary.
And sent us on our way,
Patched up and healed.

Now, quite suddenly, we call on you,
And the call is an urgent one,
You are there, of course, it never
Occurred to you that you would be
Anywhere else than at our side.
Hour after hour, day after day,
You are there, the support
Of those hands that first delivered us
Embracing us once again,
With the same love, the quiet

And gentle care; once again
We spell out our debt, our gratitude,
You say, 'It's what we do'
We nod and say, 'We always knew.'

ACKNOWLEDGEMENTS

To Anna Valentine, without whom the book would never have happened, or if it had somehow happened, would have been terrible.

To James, who did far more work on this book than I did.

To the contributors, for their time, effort, energy and brilliant writing, and without whom this book would have been very short.

To my extremely forbearing agents and cheerleaders-in-chief, Cath Summerhayes and Jess Cooper.

To every single friend and colleague who forwarded emails and begged and cajoled on my behalf, turning what should have been a Herculean task into a total doddle. Especial thanks to Jonathan Ross, Konnie Huq, Shane Allen, Chris Sussman, Jonny Geller, Jonathan Lloyd, Felicity Blunt, Debi Allen, Maria McErlane, Oliver Butler, Derren Litten, Vivienne Clore, Karolina Sutton, Jo Unwin, Sheila Crowley, Rick Hughes, Nina Gold, Louise Moore, Sarah Hitchcock, Jemma Rodgers, Andy Nyman, Pete Jones, Bev Dixon, Gordon Wise, Owen Bell and Jon Plowman.

To every single agent, manager, editor and publisher who helped make it all happen.

To the promising young cover artist and all at Pest Control.

To everyone at Orion and Trapeze listed overleaf, but with special mention to the wonderful Katie Espiner, David Shelley and Tom Noble.

To PR gurus Dusty Miller and Maura Wilding.

To Clays for their generous discount on printing costs.

And, most importantly, to everyone I've forgotten.

CREDITS

Trapeze would like to thank everyone at Orion who worked on the publication of *Dear NHS*

EDITORIAL
James Kay
Anna Valentine
Shyam Kumar
Jane Hughes

COPY-EDITOR
Liz Marvin

PROOFREADER
Kim Bishop

LEGAL
Meryl Evans
Louise Hayman

AUDIO
Paul Stark
Amber Bates
Dan Jones

CONTRACTS
Paul Bulos
Ellie Bowker
Jake Alderson

PRODUCTION
Katie Horrocks

DESIGN
Lucie Stericker
Joanna Ridley
Helen Ewing
Keith Baker
Bryony Clark

FINANCE
Jennifer Muchan
Jasdip Nandra
Sue Baker

MARKETING
Tom Noble

PUBLICITY
Dusty Miller
Maura Wilding

SALES
Laura Fletcher
Victoria Laws
Esther Waters
Frances Doyle
Georgina Cutler
Jack Hallam
Barbara Ronan
Dominic Smith
Deborah Deyong
Lauren Buck
Maggy Park

OPERATIONS
Jo Jacobs
Sharon Willis
Lisa Pryde
Lucy Brem

RIGHTS
Susan Howe

MANAGEMENT
David Shelley
Katie Espiner
Sarah Benton